Charles Howard Aaron

A Practical Treatise on Testing and Working Silver Ores

Charles Howard Aaron

A Practical Treatise on Testing and Working Silver Ores

ISBN/EAN: 9783743392656

Manufactured in Europe, USA, Canada, Australia, Japa

Cover: Foto ©berggeist007 / pixelio.de

Manufactured and distributed by brebook publishing software (www.brebook.com)

Charles Howard Aaron

A Practical Treatise on Testing and Working Silver Ores

A PRACTICAL TREATISE

ON

TESTING AND WORKING

SILVER ORES.

BY CHAS. H. AARON.

PUBLISHED AND SOLD BY DEWEY & CO.,
Proprietors of the MINING AND SCIENTIFIC PRESS, San Francisco, Cal.
1876.

Entered according to Act of Congress, in the office of the Librarian of Congress, at Washington, D. C., in the year 1876, by
CHAS. H. AARON and DEWEY & Co.

Printed by SPAULDING & BARTO,
414 Clay St., San Francisco.

PREFACE.

In 1869, I wrote a little pamphlet on amalgamation of silver ores, in which was set forth a method of treatment by which certain ores of silver usually considered refractory, could be worked to 90 per cent. of the assay without roasting. I shortly afterward introduced that method at Benton, in Mono County, with such success that two mills, of ten and five stamps, respectively, have since been built to work it, in each case with excellent results.

I myself worked the process for several years, with no better crushing machinery than an arastra, which, with an amalgamating barrel and separator (the whole driven by a water-wheel) formed a cheap and efficient mill for the purpose, though better adapted to the use of a person having a mine of his own, than for custom work.

A large portion of the matter of my pamphlet, which was sold for 25 cents per copy, has lately been published, in a slightly garbled form, as a circular, with the title of "Instructions for testing and milling ores," by one N. L. Turner, in Montana Territory, being put forth as Mr. Turner's own production, and sold, as I am informed, at the very modest price of ten dollars per copy.

In addition to the matter taken from my pamphlet, this circular contains sundry passages from articles written by me, and published in the MINING AND SCIENTIFIC PRESS, showing that Mr. Turner has been a diligent and appreciative reader of my writings, and I must give give him credit for having been so much smarter than myself, that if he has sold even a few copies at ten dollars each, he has made more out of them than I ever did.

Since writing the pamphlet, I have had five years experience and observation of the process therein described, and as my profit has consisted in part of the additional knowledge thus gained, I now propose to offer to the mining public, more and better information than is contained in that, or in Mr. Turner's piratical circular, adding also some information and suggestions in regard to other processes.

I do not write for students of science, nor to make a display of learning, which, for all the reader would know, might or might not be "all out of my own head." My aim is simply to contribute my quota to the diffusion of practical knowledge of some matters pertaining to the development of silver mining, and to make certain suggestions, of the value of which I leave the reader to judge.

I shall, then, try to write so as to be understood by common miners and prospectors, and what I say will be chiefly the result of my own experience and my own thought; not stereotyped matter which may be found in a dozen works on Metallurgy in any of our libraries.

In all silver regions there is found more or less silver ore in the form of small veins, or threads as the Mexicans say, or in bunches, pockets and deposits of little extent, which, while they will not justify the attention of capitalists, might yet furnish profitable occupation to a number of miners, if the owners only had sufficient knowledge to extract the silver in a cheap and simple way.

This want is partly met by custom mills, but aside from the fact that miners generally have a strong impression that they are very often swindled at those places, it often happens that the small and rich mines spoken of are quite remote from any mill.

In the mineral districts of Mexico, nearly every miner has some knowledge, however rude, of metallurgical operations, which enables him to work, in one way or other, any small rich "hilo" which he may discover, and though in large operations the Mexicans may not be able to compete with more enterprising people, yet it is a fact, that among miners and prospectors, a Mexican will make a good living where an American would starve to death.

Though I write mainly for the benefit of the poor and unlearned class in our mining districts, it must not be supposed that the methods of which I treat are not adapted to large operations; on the contrary, if my modification of the old "Fondon" of Alonzo Barba had been adopted for the Comstock mines long ago, many millions of dollars which are now in the Carson river, might be in the pockets of stockholders, and the value of stocks would be proportionally higher. I do not mean to say that I am the only man who could have done this, but I speak for myself, and others have the same privilege.

If it be asked why I have not tried to introduce my process at Virginia, I can only say that I have not kept the matter secret, and it is the business of those interested to see to the working of their ores in the best manner.

I have to presuppose a certain amount of knowledge on the part of the reader, in regard to the general operations of milling; for to go into all the details of feeding battery, charging pans or barrels, cleaning up, pressing amalgam, retorting, etc. would carry me far beyond my proposed limit. Most men in the mines either have, or can easily acquire, such knowledge, and my object is to enable them to use it for their own benefit, while at the same time offering some suggestions to persons of more advanced pretentions who may choose to avail themselves thereof.

C. H. AARON.

TESTING ORES FOR SILVER.

[1.] I shall not attempt to give any instructions as to the geological or lithological formations in which the discovery of silver may be expected, for several reasons. In the first place, such knowledge as we have on this subject, may be obtained from many standard works; in the second place, it is not very easliy applied when obtained, as, aside from the difficulty of at once determining the character of rock formation, a region consisting mainly of non-metalliferous rock, which might discourage a learned person from searching, may nevertheless contain within it a tract rich in metals; and in the third place, such knowledge is not altogether reliable, nor is the ordinary prospector qualified to avail himself of it, or indeed likely to be guided by anything but his own notions, derived from his own, or his friends' experience.

I shall, then, suppose that the seeker after mineral wealth, having journeyed at his "own sweet will" over desert valley and rugged mountain, has at last found a vein of ore, or at least some "rich looking float," and having brought a sample down to his camp by the spring, is anxious to know, before going further, whether the "queer looking stuff" contains silver, or only a base conglomeration of, to him, worthless metals.

Now, I am not going to afflict this innocent and much suffering prospector with scientific details, touching the crystalline form, chemical composition, fracture, streak, and specific gravity of the many different ores of silver, or minerals which carry more or less silver, in regard to which particulars he is most likely either profoundly indifferent, or profanely contemptuous. All he wants to know, for the present at least, is whether or not the ore contains silver, which he may ascertain with certainty by proceeding as follows:

[2.] Grind a few ounces of the ore to powder between two rocks; add to it about one-tenth as much salt and about half that quantity of sulphate of iron, often called copperas; mix all together, and put it into an old shovel or frying pan, which should have been previously smeared with clay or mud, and dried; then roast it over a fire, being careful to stir it often with a piece of stout iron wire. Let the roasting proceed quite gently so long as a smell of burning sulphur can be percevied, not allowing the heat to exceed a dark red as seen at night. When the fumes of sulphur cease, let the heat increase to a rather light red, but not so as to melt the ore, stirring it still with the wire. The smell will now be that of chlorides, rather pungent, often sweet as of new hay, but very easily distinguished from that of sulphur. The ore will swell, and appear woolly and somewhat sticky, and a few minutes of this hotter roasting will finish it well enough for your purpose. Now transfer the roasted ore to a flat rock and let it cool; add a little more salt, and enough water to make it like mortar; imbed in the mass a strip of clean sheet copper, and let it remain ten minutes; take the copper out, and without touching the part that has been in the pulp, wash the mud off it with clean water. If the ore contains silver, it will invariably show as a white coating on the copper, and as no other metal will so coat the copper under these conditions, the appearance of such a coating is proof positive of the presence of silver. The white coating will be heavier or lighter according to the richness of the ore, if very heavy it will appear grey and rough.

For making the above test you will require —
Salt,
Sulphate of Iron,
An old shovel or frying pan,
A piece of stout iron wire,
A strip of sheet copper six inches long,

By making a few tests in this way with ore which is known to contain silver, confidence will be acquired.

The Sulphate of Iron is not required if the ore contains sulphurets, but as you may possibly be deceived on this point by substances resembling sulphurets, it is best to make sure by using it in every case.

[3.] Many, in fact most silver ores, will indicate the presence of silver by simply mixing about an ounce with boiling water, salt and a little bluestone, in a teacup, and placing the strip of copper in it for a few minutes, while keeping it at a boiling heat; and as this is simpler than the roasting, it is well to try it first. If the copper becomes whitened you may be sure you have silver, unless indeed, it should be quicksilver, which however, you will hardly meet with in the silver regions, and which is easily known by slightly heating the copper in the fire, when quicksilver will be driven off, but silver will not. Most miners will find no difficulty in distinguishing between the two by simply rubbing with the finger.

For this test you will require—

    Salt,

    Bluestone,

    A strip of Copper and

    A teacup,

or a basin which can be set in the top of an empty oyster can, in which a little water is kept boiling, so that the basin with its contents is kept hot. I have tried this test with many samples of silver ore, and it has never failed, yet the first one is still more certain.

[4.] Some prospectors test ore by heating a small piece red hot, and then plunging it into water, when the more fusible metals it may contain appear on the surface in the form of globules. This method is not reliable, because other metals besides silver may show themselves, and either be mistaken for silver where it does not exist, or disguise it where it does.

[5.] The most generally known method of testing for silver, is by acting upon a little ore powder in a test tube with nitric acid, and adding salt to the clear solution thus formed; if silver is present it appears as a more or less dense white cloud, or, if in very small quantity, as a slight milkiness in the acid. This method is open to two objections. In the first place, some of the richest ores of silver are not acted on by nitric acid, and would therefore show no sign on addition of salt; and in the second place, lead gives the same appearance as silver, as does also quicksilver, and though an expert would easily distinguish one from the other, by further tests, yet these defects render the method unsuitable for the common prospector, besides which, nitric acid is not a very nice thing to pack with provisions and blankets, nor to handle in the desert.

[6.] I say nothing about the test with the blowpipe, because, though extremely valuable in the hands of an expert, the use of this instrument requires great practice, and considerable skill in manipulation, and the big fingers of miners do not take kindly to such delicate work. Nor do I describe the usual methods of fire assay, because they require an outfit which could not well be carried about by the prospector; nor are these things necessary till it is required to know the richness of the ore, when it is best to apply to a regular assayer, though, when it comes to that, the blowpipe assay of a carefully taken sample, or of mere specimens, is quite accurate enough, if well made, for all the purposes of the prospector, or for assorting ores, and as it costs a mere trifle, I think assayers ought to use it more than they do, as a cheap and lucrative method of supplying the wants of poor laboring men, who merely require to know approximately, the value of their ore. I have made hundreds of these assays for fifty cents each in the mines, as it takes only from fifteen to twenty minutes to make one.

TESTING FOR A PROCESS.

[7.] When ore has been found, which, by either of the foregoing tests, gives evidence of containing silver, it will be well to make some exploration to ascertain if it exists in considerable quantity, in which case a few fair samples should be taken to a reliable assayer in order that its richness may be known.

If the ore is very rich, it will, of course, bear transportation to a market; yet the losses, expenses and discounts which are inseparable from this way of disposing of it, not to mention the trouble and risk of fraud, have caused miners in general to entertain a strong, and not unfounded objection to selling their ore, especially when the market is very distant; besides which, the quantity of such ore is usually quite small, though there may be considerable of a grade which would pay quite well to work in a small mill near the mine, and owned by the miner,

Of course large mills work more cheaply, in proportion, provided they are fully supplied with ore; but a man who has only a small mine must be content with a small mill, and as the extraction of ore also costs more in this case, the profitable limit of value will be higher than when the mine and mill are large, and must be decided according to circumstances. But whether the mill is to be large or small, it is equally necessary to find out, before building it, whether the ore can be worked raw, or must be roasted or smelted.

[8.] If fuel is abundant, smelting is preferable when the ore contains a large percentage of lead, and becomes indispensable if the quantity is excessive. However, ore containing as much as thirty per cent. of lead can be worked by roasting. If not necessarily a smelting ore, it remains to be seen whether it can be worked without roasting or not, which can be ascertained as follows, so far at least as processes which I can recommend to unlearned persons are concerned:

[9.] A fair average sample is taken and reduced to powder, fine enough to pass through at least a

40-mesh wire gauze. If, however, the sample is large, it will suffice if it be crushed coarsely at first, then, after thorough mixing, spread upon a smooth floor and halved; one-half being rejected, the other is crushed finer and again mixed and halved, and so on until the last half equals about six pounds, which is all passed through a fine sieve. Of this fine pulp, again well mixed, five pounds are carefully weighed for a working sample, and the remainder set aside for an assay.

[10.] The working sample is put into a porcelain-lined iron kettle, such as are used in a mill for carrying quicksilver; water enough to form a thin pulp, and two or more ounces of salt are added. The pot is then put on a stove or over a fire, and heated to boiling of the pulp, while stirred with a strip of wood to prevent the ore from settling to the bottom, which would cause the destruction of the enamel. When hot, a little strong solution of bluestone is added to the pulp, and a piece of sheet copper about five inches square, or several smaller pieces, together with about four ounces of quicksilver, are also put in. As the pulp will gradually dry by evaporation, hot water is added from time to time, and the whole is frequently stirred with the strip of wood.

After from five to ten minutes, a clean strip of iron, or the blade of a case knife, is dipped into the pulp for a few seconds, then washed, without rubbing, in clean water; it should show a pretty strong color of copper; if it does not, more solution of bluestone is added to the pulp, and the trial repeated, and so on till the iron becomes coated with copper, when it is known that enough bluestone has been used; the heat and stirring are continued, the sheet copper in the pot being pushed about through the pulp.

It is absolutely necessary that there be copper in solution in the pulp, and as there may be in the ore some substance which will destroy a quantity of the bluestone used, it is best to repeat the test with the knife blade once or twice more, adding bluestone if necessary, till it is found that a permanent reaction of copper is established; but the stirring must be done with a stick, not with the iron knife, as that is only used as a test to show if the pulp is in proper condition. The continued presence of iron would ruin the experiment.

After a short time the pieces of sheet copper will

become coated with quicksilver, and will appear rough on the surface, from adhesion of silver amalgam and corrosion of the copper, and particles of ore are seen to adhere slightly to the copper and to the quicksilver, and this, whether in the test or in the large working, is an indication that the operation is proceeding favorably, while the cessation of the phenomenon is an indication of its completion.

The operation is continued till a clean strip of copper dipped into the hot pulp for several minutes, comes out perfectly free from a white deposit of silver, which may require from three to ten hours.

[11.] When about to use the copper test, the pulp is made rather thin, by the addition of hot water, to allow the quicksilver to settle to the bottom, and care must be used not to touch it with the copper. If, notwithstanding all precautions, the copper should appear spotted with quicksilver, it must be heated in the fire, scoured clean, and the trial repeated. The iron test is used at the same time, to make sure that the pulp is in the proper condition, that is, that it contains a solution of copper.

In commencing this operation, it sometimes happens that the iron tester is blackened, instead of being coppered; in this case the work is continued some little time without addition of more bluestone, when, on repeating the trial, the proper effect will usually be got.

[12.] In order to know how much bluestone has been used, it is best, before beginning, to weigh two ounces, and put it into a bottle which will hold enough water to dissolve it, adding one ounce of salt, and filling the bottle with water. If, at the end of the operation, the bottle is half emptied, it is clear that one ounce of the bluestone has been used, and so in proportion for less quantities. It is important to know how much bluestone has been used, as calculation may then be made of the expense of working on the large scale; thus, if 5 lbs., or 80 ounces of ore consumes 0.5oz. of bluestone, then one ton, or 2,000 lbs., will require—

$$80:0.5::2000:12.5 \text{ pounds.}$$

Some ores require more, but the Comstock ore works well with 4 pounds to the ton. The salt used on the large scale will be from 50 to 100 lbs.

to the ton of ore, and in addition, one-half the weight of the bluestone.

[13.] When the strip of copper shows that no more silver can be got, although the pulp is boiling hot, and the case knife shows the continued presence of a solution of copper, the adhering amalgam is removed from the sheet copper, by stripping it between the fingers and thumb, the pulp is "panned out" of the pot, and the amalgam strained out of the quicksilver with a piece of buckskin or wet drilling. The ball of amalgam thus obtained, which will be white and crisp, is tied up in a piece of rag, and placed in an assay crucible, or any vessel which will stand heat, and is slowly heated to redness. After cooling, the spongy silver is assayed to ascertain its value, which, multiplied by 400, gives the amount per ton that the ore will yield. If the silver is clean, as it ought to be when worked in this way, it may be estimated, without assaying, at $1.25 per Troy ounce.

For making this test, the following appliances are required—

A pestle and mortar (or a flat rock and small boulder);

A sieve of at least 40 meshes to the linear inch;

A porcelain-lined kettle;

A strip of sheet copper, and some scraps of the same metal;

A case knife;

Quicksilver, salt and bluestone;

A strip of wood for stirring, and patience.

Persons who are accustomed to such operations can make this test on 1000 grains of ore, and, to such, the following statement may be more acceptable than the foregoing details: A weighed quantity of the pulverized ore is digested in a glass or porcelain vessel, with a solution sulphate of copper and salt, metallic copper, and quicksilver, and the resulting amalgam is estimated proportionally for a ton of ore. If a fire assay is made of the unworked residue of the sample, a comparison of the two results will indicate the working percentage.

[14.] Almost any ore will yield a portion of its silver to this treatment, and all those which give a good result, can be worked in the large way by Aaron's process; some by Patchen's process, and a few by simple pan treatment. Ores which will not yield to it cannot be treated successfully by any of these methods, but will in general be best worked by roasting, and I know of none in this country which cannot be so worked successfully.

The test by roasting cannot be made reliably on a very small scale; but this is not necessary, as it may be taken for granted, that, setting aside smelting ores, roasting will give a fair result if rightly managed.

WORKING ORES.

[15] Aaron's process is essentially similar to the working test just described. I have called it Aaron's process because I invented it, independently, in 1867, being led to it by considering the reactions of the Patio, and because I have reduced it to a fair degree of practicability. My first operation was conducted on about a grain of ore in a minute porcelain cup, with the aid of a copper belt-rivet; the next was on five pounds of ore in a kettle; then on a ton in a wooden barrel, and subsequently, thousands of tons have been worked by it, and near a million dollars extracted. I soon found that the process was similar to the Mexican *fondon*, which, however, never advanced beyond a very rude and primitive stage in the hands of the Mexicans, and was in that form totally unadapted to our day and country. The loss of quicksilver was enormous, except when the ores consisted of chlorides, bromides, and iodides, which do not require

the use of bluestone or *magistral*, being readily worked in copper or iron pans, with salt and quicksilver alone. My process, as now worked, involves no greater loss of quicksilver than other methods of amalgamation in barrels or pans, but rather less, and the chemical and mechanical treatment differ materially from the old *fondon*.

[16.] The ore, crushed wet or dry, and at least fine enough to pass through a 40-mesh wire cloth, is put into Aaron's amalgamator, or a wooden amalgamating barrel, together with a quantity of metallic copper, 50 or 100 pounds of salt, and enough water to form a pulp of medium consistency. A certain quantity of dichloride of copper is added, the barrel closed and put in motion, and its contents heated by means of steam admitted through the journal. When the pulp is boiling hot, a little of it is taken in a small porcelain-lined sheet iron saucepan, and tested for copper in solution, with the blade of a case knife, as already described. If no copper is found, more dichloride is put in, and the test repeated after the barrel has been again in motion some little time.

[17.] When the presence of a solution of copper

is indicated by a deposit on the knife blade, (remembering the remark about a black stain), quicksilver is put in, and the barrel is kept in motion from 6 to 12 hours, during which it will not usually require more steaming, and being again stopped, a little pulp is taken as before, diluted with some hot water, adding a little salt to maintain the strength of the brine, and is tested, while kept boiling hot with a few live coals, for silver, with a strip of sheet copper, just as directed for making the working test. When no silver can be found, the barrel is discharged into the separator and recharged as before.

[18.] In treating ores raw, when crushed wet, some inconvenience results from the condensation of the steam used for heating, by which the pulp is too much diluted. This may be remedied by using superheated steam, whereby not only is the dilution avoided, but the pulp may even be thickened if necessary. If the ore is crushed dry, a small portion of the charge is kept back, till the contents of the barrel, or other machine, are heated, when a quantity sufficient to make the pulp of the desired consistency is put in.

[19.] The dichloride of copper is prepared as follows: 22 pounds of bluestone and 10 pounds of salt, dissolved in water, are boiled, together with five pounds of iron borings, in a steam heated rotating barrel, or a vat, till the iron, and copper precipitated by it, have quite, or almost wholly disappeared, being replaced by a light colored powder, which is dichloride of copper. Or, 11 pounds of bluestone and 5 pounds of salt are boiled, together with a quantity of scrap copper, till the solution is no longer green, when the same quantity of dichloride of copper is formed. If the salt is impure, a larger proportion must be employed; an excess does no harm, but in no case should the liquor be thrown away, as it may contain some dissolved copper, so that it, as well as the precipitate, should be used. The dichloride of copper should not be much exposed to the air, as it thereby becomes changed.

[20.] Protochloride of copper, most conveniently prepared by dissolving some bluestone, with half its weight of salt, in water, may be used in place of dichloride, and in precisely the same manner as to tests, etc., but, as it attacks quicksilver, the following precautions are necessary: The quicksilver must not be put in till some hours after the pulp has been heated and worked with the copper balls, so that the chloride of copper has lost its power of acting on it, and then, in order that there may be no quicksilver in the amalgamator, or sticking to the copper when the new charge of ore is put in, it must be used very sparingly, not more than twice the weight of the silver in the ore being used, or about 3 pounds to $100. The amalgam is then found mixed through the pulp in the form of a coarse, gray powder, or little flakes, and must be collected by working the pulp in a wooden pan, with an additional quantity of quicksilver, and slightly diluted with water, so that it may be drawn off, leaving the bulk of the fine amalgam in the pan, to the separator, where it is treated for recovery of copper and quicksilver, as described hereafter.

[21.] Ores containing carbonate of copper do not require the dichloride prepared as above. Instead of it sulphate of iron is used, and is put at once into the barrel with the ore and salt; it produces the dichloride and protochloride of copper from the carbonate in the ore. ⟨Sulphurous acid is better than sulphate of iron for this purpose, as it forms only the dichloride of copper, besides having other advantages, and being much cheaper, for as it is only the sulphuric acid in the sulphate of iron which is useful, one pound of sulphur in the form of sulphurous acid, is equal to about 10 pounds of sulphate of iron.⟩ The sulphurous acid is made by burning sulphur, and the vapors are either forced into the wet ore-pulp, or condensed in water, with which the dry ore powder is afterwards mixed. I have a patent on this reagent for the purpose of amalgamation.

[22.] I have often been asked, when working this class of ores, why I used the copper balls in the barrel when there was so much copper in the ore. The reason is that the process requires copper in solution, and metallic copper. Now the copper in the ore is not metallic; it is not really copper, it is carbonate of copper, just as iron rust, though formed from iron, will not do to make a crowbar of, so in our operation the balls of copper must be retained. Still the copper contained in such ores, can be and is utilized as follows: a certain quantity of iron borings is put into the barrel, together with the charge of ore, salt and sulphate of iron, and by a complex reaction, which I shall not explain here, economizes the consumption of the balls by producing copper from the ore, without, however, diminishing the quantity in solution. The quantity of iron to be used in this way depends on several circumstances, and varies from 2 pounds to 50 pounds to the ton of ore. Rich ore will bear more than that which is poor, but it must be regulated by the quality of the bullion, as too much produces a coppery amalgam. It is not essential to use it at all, but it is economical. Iron is not used in this process except when working ores containing carbonate of copper, or as already shown, in preparing the dichloride, and as the quantity must in all cases be under control, the process cannot be worked in iron pans.

[23.] The quantity of chemicals, iron, etc., required in working any given class of ore is soon found, so that after working a couple of charges there is no need of testing every time. In Benton the average quantity of sulphate of iron used was 22 pounds per ton of ore.

[24.] Carbonate of lime and of lead, if in great quantity, make these processes expensive by consuming too much of the chemicals, and can only be overcome cheaply, by means of sulphurous acid, which, moreover, has many advantages over sulphate of iron in working coppery ores.

[25.] Aaron's process has the great advantage of always giving fine bullion, no matter how base may be the ore, if properly managed, and the quicksilver always comes out clean. Among the ores which can be worked by it are partzite, stetefeldtite, silver glance, carbonate of silver, ruby silver, etc.; galena, zincblende and copper pyrites, containing silver, do not yield well.

[26.] The ores of this country are usually very mixed, and often it is difficult to determine in what condition or combination the silver exists in them, so it is not always safe to say what process will prove best, without making a trial. Some ores which prove refractory in the raw state, may be made workable by roasting in open or covered heaps or kilns. It is easily tried by burning a small pile and then taking a fair sample and mak-

SILVER ORES.

ing a test on 5 pounds. Chloride ores, as already stated, require no special treatment, being readily worked in barrels or pans with salt and quicksilver; but I wish to observe that of the ores commonly called "chlorides" by the miners, not one in a hundred contains any chloride, being, in nearly every case, simply an ochreous compound of base metal, which may or may not carry some silver.

[27.] A separator with an iron bottom which is scoured by wooden muller shoes, is used in connection with this process, by which means a large loss of quicksilver is avoided. The machine I have found to answer best is described further on. In it the warm pulp as it comes from the barrel, containing dissolved copper, is worked, without addition of water, for a greater or less length of time, whereby a quantity of copper amalgam is formed on the iron bottom, and nearly all the mercury is saved. On this account the bulk of quicksilver and amalgam is not allowed to pass to the separator, where it would be mixed with the base amalgam there formed, but is retained in the trough through which the pulp flows, by means of a riffle, or "pot," and after being sponged clean, is led by an iron pipe to the strainer.

The base amalgam from the separator may be disposed of in several ways—

1st. It is returned to the barrel, where it answers the same purpose as metallic copper, and saves the consumption of the balls, but too much will spoil the bullion.

2d. It is retorted, and as in this operation there occurs a partial separation of the copper from the silver, of which latter there is always more or less present, the spongy metal is broken to pieces, and assorted into two or three grades; the finer part is melted and sold, and the nearly pure copper is used to make the dichloride for future use.

3d. The retorted metal is melted, and cast into bars or balls for use in the barrel.

Thus the metallic copper consumed in the amalgamator, and with some ores that also which is used as dichloride or bluestone, is recovered at the expense of the iron bottom of the separator, and in the case of ores containing carbonate of copper, as much copper as is desired may be made so long as there is any left in the ore to get out. It would be perfectly possible to work copper ore for copper on this plan, but we can do better by leaching on the Hunt and Douglass plan.

In working the ores of Mono County, it was found that when a separator made entirely of wood was used, a large quantity of quicksilver passed off in the form of a red powder, easily detected by panning, but containing no visible metallic quicksilver; the use of the iron bottom prevented this loss by decomposing the red powder, but in order to produce this effect there must be copper in solution in the pulp, which however is always the case if it has been properly worked in the barrel. After a sufficient time the separator is filled with water, and subsequently discharged in the usual way by plug-holes. Amalgamators specially adapted to this process, will be described in their place.

PATCHEN'S PROCESS.

[28.] The Patio is the mother of Aaron's process, the Fondon is its elder sister, and Patchen's process is a younger member of the same family. As it is patented, a minute description is not needed here. It consists essentially in treating the pulverized ore with dichloride of copper, and without quicksilver, in a copper-lined, steam-heated vessel, after which it is transferred to an iron pan, and worked with quicksilver. The process commends itself to Virginians by making use of the iron pans with which all their mills are provided, but equally good results are got by Aaron's process, and as the tests described for them are not applicable to this, the operation is not so easily regulated, whilst in some cases the bullion obtained is not fine, owing to the action of the iron pan in extracting copper or lead.

ROASTED ORES.

[29.] Roasted ores may be worked in barrels with scrap iron or copper balls, or in iron pans. Ores containing much copper, lead, etc., will not yield fine bullion in pans, unless the roasted ore is first washed in filter vats with acidulated water, to dissolve out the salt which may remain after roasting, and the base metal chlorides, some of which are not soluble in plain water. Barrels with scrap iron give a finer bullion from such ores, than pans, but any roasted ore will yield fine bullion if worked in barrels with copper balls, or in the new machine hereafter described. When these machines or barrels are used, it is sometimes necessary to sift the ore to remove lumps formed in the furnace, which are either returned to the battery or ground in a pan. On this account, pans are usually preferred, though the wisdom of the choice is doubtful, especially if the ore is base. I have in some cases got 90 per cent. of the silver from roasted ore, in barrels, without sifting. As nearly all the salt used in roasting is decomposed or evaporated, it is very advantageous to add a little to the pulp in the amalgamator; it acts by dissolving the chloride of silver, which aids its reduction to metal. . The quality of bullion obtained from base roasted ore, when worked in iron pans, may be improved by adding a certain quantity of lime or ashes, but an excess must be avoided.

[30.] It is a common error to suppose that the object of roasting is to "destroy" the base metals, and this absurdity has been encouraged by some who ought to know better; hence many think that roasted ore ought always to yield fine bullion. The truth is, that silver ore is roasted with salt for the purpose of changing the various and refractory compounds of silver into one known and easily reduced compound, namely, chloride. Now if this could be done without affecting the base metals at all, we should always get fine bullion, because the base metals would not be extracted by the pan; but it unfortunately happens that the roasting, which aids us to get the silver, puts the copper, lead, etc. into a similar condition, and they are not driven off to any great extent in the operation. It is true that, by a certain modification of the roasting process, the base metals can again, in a measure, be changed so as not to be amalgamated, but this causes loss and expense, and is never fully effective, besides which the chemical action in the pan is such as first to reproduce and then decompose the base metal chlorides, so that after all they find their way into the amalgam more or less; nor can this action be entirely prevented by the use of potash or lime in the pan, since the quantity of these which would produce the required effect, would also interfere with the amalgamation of silver.

[31.] The chloridizing roasting of silver ore is not so mysterious or difficult an operation as some suppose, and others pretend—that is, practically. The proof of this is the success attained by many persons of no scientific attainments, and of quite moderate ability. What is principally necessary, is courage to try, and then courage to try again.

Of course in a custom mill, where different kinds of ore are constantly coming in, and where no failures are allowable, knowledge and experience are indispensable in the operator; but any person of common intelligence may soon learn to roast a given kind of ore, especially when he is his own boss, with no one to grumble at his mistakes. In large operations the mechanical roasters, such as Stetefeldt's, White's or Brückner's, should be used; but for a small mill, working about two tons of ore per day, and owned by the miner himself, a reverberatory, such as I shall describe, will answer perfectly, though if one can afford to, get a Brückner and run it, say one or two days in a week, the work will be done cheaper.

[32.] To roast ores well requires practice, the directions I shall give, will therefore be brief and general in their character, to be interpreted and applied with judgment. A knowledge of the theory should be indispensable to the professional man, who ought to know the *why*, as well as the *how*, to do things; but our "honest miner" only wants to know how to get a fair percentage of the silver out of his particular ore.

[33.] The essentials of chloridizing roasting are heat, air, sulphur and salt. Heat is obtained by means of fire; air enters the furnace through the fire-place, and through openings in the sides; salt is supplied by the operator, and sulphur either exists naturally in the ore, or is added to it in some shape. Ores having lime in them, require more sulphur than such as have a quartz gangue.

[34.] Night is the best time to study roasting, as the different degrees of heat are then most readily seen, and for the same reason the furnace is best in a place where there is not too much light in daytime. What I say about the heat must be considered as referring to appearances as seen at night, and when the fire is not flaming.

[35.] The furnace is heated to a dark red, the ore is put in, spread on the hearth, and stirred at short intervals with hoe or rake. From 5 to 10 per cent. of salt is charged with the ore. It is best to begin with the larger quantity, and when the roasting is found to yield a good result, less salt may be tried, till the lowest quantity that will answer is found. If, on stirring the ore, the sulphur is seen to burn with a blue flame, the fire is kept very low, and the ore stirred continually till no more sulphur is seen to burn, which may, in extreme cases, take several hours, during which the fire is allowed to almost go out. The heat is then raised slowly to about a cherry red, the ore still being stirred, and not till a sample, taken from the furnace with a drill spoon, smells no longer of burning sulphur, is the heat raised, gradually, to a light red, approaching to white, but not so as to slag the ore, which at this time assumes a condition which has been described as woolly, spongy or feathery, and swells greatly, while, whereas at an earlier stage it appeared very mobile, running and spreading like water, it is now rather sticky, and "stays put." As at this time less constant stirring is needed, this is a good opportunity to scrape the ore thoroughly out of all the corners, piling it in the middle of the hearth, after which it is again spread. When all the ore is equally heated, so that on stirring clear to the bottom, and in the corners, no black places are seen, the fire is allowed to go down, ready for another charge, the ore being drawn away from the bridge, and moved only enough to prevent any part from getting so hot as to melt. It is well to draw it into a heap ready for discharging, which is done as the furnace cools off.

[36.] On taking a sample of the roasted ore, and panning it down in a saucer, no raw sulphurets should be seen. Lumps, more or less, there will be, but there should be no slagged or sintured ones, except as occasionally, by accident, a little ore has stuck to the bricks, and after being slightly melted, has been scraped off again; and right here I want to say something about stirring. Never push the hoe so as to jam the ore against the side of the furnace, but when it is within about a foot of the bricks, lift it clear of the ore by lowering the handle, push it on over the ore, and then let it down and move it the other way. Thus you will avoid plastering the sticky ore on to the bricks, and save much trouble.

[37.] The chief danger in roasting is in applying too high a heat at first, but ores which contain little sulphur may be heated more rapidly than those which have much. Some ores with calcareous gangue roast to 90 per cent. without at any time smelling strongly of sulphur or of chlorides. The color of the roasted ore varies greatly, depending on the constituents, the degree of heat to which it is exposed, and the exposure to air, both in the furnace and after it is withdrawn. The knowing just when to discharge the ore is a matter of judgment and experience, the only reliable test of its being well roasted, except by working it, being such as to require a knowledge of assaying. There is however, a test, in addition to that by washing in a saucer already spoken of, which may afford a clue. It consists in throwing a little of the red hot ore into a dish of water; well roasted ore will diffuse a white cloud in the water, whilst if the roasting is not finished, the cloud will be more or less yellow.

[38.] Ore which fails to roast well for want of sulphur, requires the addition of 2 or 3 per cent. of iron pyrites, or sulphate of iron, or a little pulverized sulphur.

LEACHING PROCESSES.

[39.] Leaching processes are of several kinds, for gold, silver, or copper, or all three combined.

They all involve the necessity of roasting, with or without salt. The roasted ore is put into filtering vats, and the metals are extracted, just as lye is extracted from ashes, by passing through the mass a fluid which dissolves them, after which they are separated from the fluid by the addition of a substance which throws them down, either in a metallic state, or as insoluble compounds, which must be again treated for the pure metal. These processes are very admirable, and in many cases cheaper than amalgamation. Some difficulty is frequently found in getting the dissolving fluid through the ore fast enough, and to overcome this, I have contrived two different machines, to be described hereafter.

SMELTING.

[40.] It has already been mentioned that, in certain circumstances, smelting is the best process for extraction of silver. As I have no experience in this line, any instructions I could give would only be copied from some one else, to which some one else I therefore beg to refer the reader who is anxious to smelt his ore. He can find much about it in various books, after studying which he will still have much to learn,

SILVER ORES. 48

[41.] When a Mexican finds rich ore and can get galena in the vicinity, he puts up a little furnace of adobes, smelts out his silver-lead, refines it in another little furnace, and buys his beans with the proceeds. Why cannot intelligent Americans, who have opportunities of seeing smelting carried on, go and do likewise? In course of time the knowledge would spread, and many honest miners might profit by their discoveries, instead of waiting for capital, or abandoning their mines, because they cannot sell them.

A MEXICAN PROCESS.

[42.] The terreros, containing fragments of ores, consisting of oxides and sulphides, are ground to an impalpable powder in the arrastra, with addition of the new magistral, for some hours, at the expiration of which period quicksilver is added, and the grinding continued till all the silver be amalgamated. Ores containing chloride, or iodide of silver do not yield to this process, and compounds of chlorine in every form must be carefully excluded. The grinding must be done in stone, not in iron arrastras. Excess of the magistral is to be avoided, as it vitiates the amalgam.

The new magistral is made by digesting suboxide of copper with sulphuric acid, in equivalent proportions, which forms an intimate mixture of sulphate of copper and finely divided metallic copper.

For the above description I am indebted to Mr. John Scott, of San Francisco, Assayer. I know nothing more about the process, but it is not clear to me that a mixture of cement, copper and bluestone would not answer as well as the above preparation, which must be rather expensive.

KRŒNCKE'S PROCESS.

See Berg-, Hütten-, und Salinen-Wesen 1876. V. 26. p. 487.

[43.] Krœncke's process is practiced in Copiapo, Chili, and is described as follows in the *American Chemist*:

"The ores of this district, consisting of the chloride, iodide and bromide of silver, together with various sulphides of silver, occur in Jura limestone and marl, diabase and porphyry, and have associated with them, as veinstones, calcite, heavy spar, gypsum, amianthus, kaolin, and in the upper levels, ferruginous clay. There has never been any difficulty in completely extracting the metal from the chloride, bromide and iodide by amalgamation, while the sulphurets have always been the cause of large losses in the residues. Krœncke first based his process on the treatment of these residues and subsequently extended it so as to treat the pure sulphides, arsenides and antimonides of silver. The writer made a series of experiments in order to arrive at a theory of the process. He found:

1st.—If ruby silver be digested in a finely pulverized condition with a hot concentrated solution of cupric sub-chloride and sodium chloride, a chemical reaction is soon perceptible, the powder turning black. An analysis of this powder shows the formation of argentic sulphide and cupric subsulphide, while antimonic chloride is found in the solution. The reactions are—

$$3AgS, Sb S_3 + 3(Cu_2 Cl) + NaCl = 3AgS + SbCl_3 + 3(Cu_2 S) + NaCl.$$

2d.—If the silver sulphide thus obtained be again treated in a hot solution with cupric sub-chloride and sodium chloride, and zinc is added, metallic silver is almost instantaneously formed. The reactions are—

$$AgS + Cu_2 Cl + NaCl + Zn = Ag + Cu_2 S + NaCl + ZnCl.$$

The zinc probably acts as electro-positive metal,

predisposing the atoms of argentic sulphide and cupric sub-chloride to a mutual exchange, so that the cupric sub-sulphide and argentic chloride are formed; which last is decomposed in a nascent state by the zinc, with the formation of zinc chloride and silver. If the very finely divided argentic sulphide obtained in the first experiment be taken, the action of the reagents is almost instantaneous. If, however, the experiment be made with a small piece of argentite (AgS) in contact with a small piece of zinc, a dull white coating is formed on the surface of the argentite, which is probably argentic chloride, and which only assumes a metallic appearance with prolonged action of the zinc.

3d.—If in the second experiment mercury be also added, the reduction of the silver takes place still more rapidly, while at the same time amalgam is formed. Lead can be used instead of zinc and both act more powerfully when used in the form of amalgam. The chief points in this method of treatment are:

1st.—The use of a hot concentrated solution of cupric sub-chloride and sodium chloride, especially of the latter, in order to retain a larger amount of cupric sub-chloride in solution, and to prevent the formation of basic copper salts.

2d.—The use of perfectly dry ore, finely pulverized, in order that it may easily absorb the solution. Wet ores cause the formation of basic salts, and hence loss of effective cupric sub-chloride, as well as imperfect impregnation.

3d.—The use of cupric sub-chloride and of lead or zinc in combination with mercury as amalgam, in such quantities that the reactions described may take place through all the ore without any excess of the last-mentioned metals. Since an excess of cupric sub-chloride is generally used, this last will be decomposed by the lead (or zinc) and copper formed, which pass into the amalgam and whose removal is always troublesome when present in large quantities. Kroencke treats very cuperiferous amalgam, after having been squeezed and ground in a centrifugal apparatus, with a hot solution of cupric chloride in order to extract the copper. ($CuCl + Cu = Cu_2Cl$.)

4th.—It is always advisable, so far as practicable, to treat ore of constant composition, as otherwise experiments on a small scale must be constantly made, even when but slight variations occur in the nature of the ore. The presence of mere traces of blende make the process much more expensive, while large amounts of the same render the use of this method impossible, as the cupric sub-chloride solution is instantly decomposed by the blende."

THE CHILIAN PROCESS.

[44.] The Chilian process, described in the *Mining and Scientific Press* of January 10th, 1874, is similar to Aaron's process, and presents no advantages over the latter. Other methods are treated of in standard works on the subject.

PULVERIZING MACHINES.

[45.] On this subject I shall confine my remarks to machines which are calculated for small operations. The splendid stamp batteries now used in our first-class mills call for neither description nor comment here.

THE ARRASTRA.

[46.] The arrastra is the poor man's mill. It consists of a circular pavement of stone, upon which the ore is crushed by drawing over it heavy stone drags, attached by chains to a horizontal beam, or two crossed beams, pivoted to a post in the center of the pavement, motion being imparted either by water, steam or horse-power. The stone pavement is surrounded by a low wall of rock or wood, to retain the ore and water during the grinding, and in which is a gate for discharging the pulp. The Mexicans often build an arrastra and water wheel in one, the beams to which the drags are attached being prolonged to form the spokes of a horizontal reaction wheel, surrounding the arrastra, which is driven by a stream of water from an inclined sluice.

[47.] When used for grinding silver ore, as the quicksilver is not put into the arrastra, there is no need for nicety in its construction, nor for tamping between the bottom stones. The pavement is made of rather thin stones set up edgwise on a bed of sand, and should not be laid nearer than six inches from the outer wall, or the center post, otherwise it cannot be kept flat as it wears down. The spaces thus left are filled with gravel or clay. The first charge worked in a new arrastra consists of some barren rock and clay, by which all interstices are filled, to avoid loss of valuable ore. An arrastra of 10 feet diameter, with drags as heavy as can be drawn by a 2-horse power water wheel, or about 40 miners' inches of water, on a 30 foot overshot wheel, will grind from $1\frac{1}{2}$ to 2 tons of ore in 24 hours, in charges of 1,000 to 1,500 pounds. It is placed so that the pulp, when sufficiently ground, flows directly from it to the amalgamator, which is on a lower level. The pulp is kept as thick as is consistent with good grinding.

SILVER ORES.

[48.] The arrastra may be made to grind and discharge continuously, by making part of the surrounding wall to consist of a fine screen of wire cloth, through which the pulp flows into a circular trough, and thence to a vat. Water flows in continuously, and is regulated so that while the pulp is thin enough to flow through the screen when fine, yet the sand is not allowed to settle so as to pack on the bottom. Ore is fed in at intervals of about half an hour. The water which flows from the settling vat is either returned to the arrastra by means of a wooden pump, or an archimedian screw, or it is led into a reservoir where all suspended matter is deposited. From 8 to 10 revolutions per minute is the proper speed for a 10-foot arrastra. The drags are as heavy as the power will allow, and so arranged as to keep the bottom flat as it wears; they are attached to the beams by chains, hooking into iron eyebolts set into the rock with melted lead. The lead after being poured in, shrinks in cooling, so that it is a little loose, and it is tightened by being driven down with a set or punch. The drags are so hung that the forward end is lifted a little, which prevents catching and jerking; they are also retained in the required position by stay chains at the rear ends.

[49.] The arrastra will grind dry as well as wet, the only trouble is to get the ore ore out as fast as it is ground, which is done while the machine is in motion, with a shovel, and it is sifted by a rotating cylindrical sieve, so arranged that the coarse stuff falls again into the arrastra. This is very poor work, and I think it might be done much more conveniently by making the enclosing wall to consist of a set of fine screens on a circular framework; then, by means of a sort of scoop, somewhat similar to a plowshare, attached to one of the rotating beams, the ore will be thrown forcibly against the screen, and the fine portion will pass through. Such an arrastra should be worked at 15 to 20 revolutions per minute, should have a light removable cover, and the bottom, instead of being flat, should be conical, deepest in the center, where the ore should be fed in at intervals. By this arrangement very coarse ore will never reach the screens to break them. This machine should be driven by a belt, to avoid serious damage in case of a drag getting loose.

STAMP BATTERIES.

[50.] Stamp Batteries with 4 light stamps of 400 pounds each, are preferable to those with two of 800 pounds each, because the parts are more easily handled, and the strain on the machinery is more even. In wet crushing the water should never, in a silver mill, be allowed to run off, but should be pumped, or raised with a screw, from the settling tank, and passed again to the battery. It is a good plan to saturate the water with salt, which facilitates the settling of slimes, and in cold weather, tends to prevent its freezing. Ore can be crushed nearly as fast dry as wet, provided it can be got out of the mortar as fast as it is reduced fine enough. The discharge is facilitated by the following arrangement: .

[51.] In place of the usual fine screen in the mortar, a coarse one of 8 or 10 meshes to the linear inch is used; the ore passing rapidly through, falls on to an inclined apron, formed of fine wire cloth stretched on a frame, which is hung to the battery posts by hooks and eyebolts, and by iron rods, arranged with buckle screws, by which they can be lengthened or shortened to alter the inclination as required. This apron is five or six feet wide one way, and as wide as the mortar the other way. The fine ore passes through into a receptacle below, while the coarser part passes over the apron and falls into another box. In some cases the coarse matter consists almost entirely of gangue, and is too poor to be worked, so that a certain concentration of the richer part is effected, but otherwise it is returned to the battery together with fresh ore. In large mills conveyors and elevators dispose of the coarse and fine material respectively, as may be desired, but in a small one this is done by hand. No "joggler" is required to shake the apron, as the jarring of the battery to which it is attached is sufficient. In order to be crushed fast, and pass readily through the screens, ore must be thoroughly dried by artificial heat, especially if it contains clay or talc. It may appear perfectly dry and yet contain as much as 7 per cent. of water, which will evaporate in the heat produced in crushing, and by condensing on the screens, will cause the fine dust to stick. The drying may be done in a kiln or oven, and the ore fed into the battery quite hot. Where the ore has to be burned in heaps before working, it will be crushed very easily.

CROCKER'S TRIP-HAMMER BATTERY.

[52.] Crocker's Trip-hammer Battery is a machine of which, on theoretical grounds, I think very favorably, and I can state from observation that it crushes rock very fast. The advertisement states that it crushes 600 pounds per hour, requiring one-horse power to drive it. The advantage of applying a given amount of power in the form of a light hammer, moving at high velocity,

rather than a heavy one at low velocity, is well known, and is illustrated by the breaking of stone for macadamized roads in England, which is always done with a light hammer on a long handle. In this battery the velocity of the blow is obtained by means of wooden springs, acting under the short arms of the levers to which the hammers are attached. As to the stated power required, it is probably under-estimated, as a calculation by the usual empirical formula of mechanics, based on the width and speed of belt used, gives three-horse power as the result. These calculations, however, are not exact unless the tension and material of the belt, and extent of contact are strictly taken into account; and I find that a calculation based on the number of drops, height of lift, and weight of hammers, gives but little over half a horse power as the theoretical amount required; to which must be added an unknown quantity for friction, and for the excess of weight of the longer ends of the arms over that of the shorter ends. The truth probably lies between the two estimates. Allowing, however, that it takes three-horse power, the performance of the machine shows an immense gain over that of the common battery, while its small size and weight, being only 1,500 pounds; and the fact that it requires no building up, being all ready for use when bought, and may be taken to pieces and set up by one man, combined with the low cost, entitle it to the favorable consideration of miners, and adapt it especially to the use of that class of small mine owners, to whom principally I address myself.

In working, the springs are not compressed by the direct action of the cams, but are so adjusted, that if the cam shaft be revolved slowly, no compression takes place; but at the high speed used in actual work, the acquired momentum of the rising hammers, carries the levers beyond the direct range of the cams, and causes the short ends to come in contact with the springs, by the elasticity of which they are driven upward, causing the hammer to descend with great velocity. The information I have been able to obtain at the foundry does not indicate any undue liability to breakage.

PAUL'S PULVERIZING BARREL.

[53.] Paul's Pulverizing Barrel is a happy conception, the value of which is not yet generally appreciated. It is most advantageously used to follow a battery which crushes dry through a screen of about 30 meshes to the linear inch, and is excellently adapted to the preparation of silver ores for

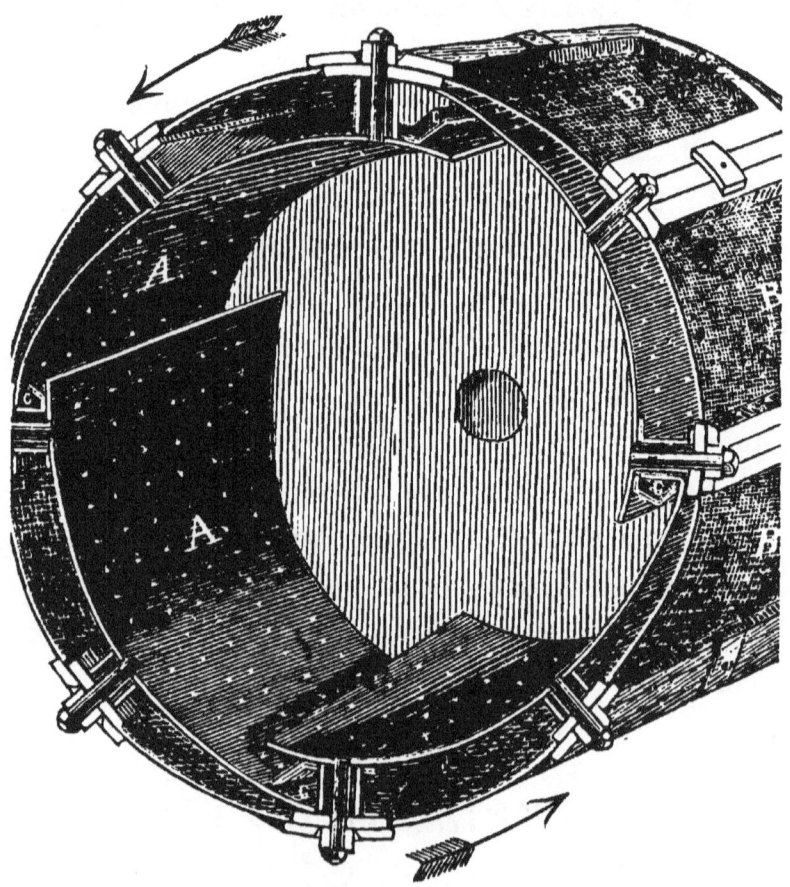

raw working, or of roasted ores for amalgamation in machines which do not grind, as barrels. As rock is used to grind rock, there is less wear of iron than in other iron pulverizers, a point which is of great advantage when treating ores by those processes which depend on the use of solutions of copper, except indeed, in those cases of copperous ores, in which, as already explained, a certain quantity of iron may be used with advantage. There is an impression among some, that this machine works very slowly, but I think this judgment has been formed without due consideration of the work done. A machine which will take 10 tons of ore, as it comes through a 30-mesh screen, and will pass it all through a screen of 120 meshes to the linear inch in 24 hours, using only 4-horse power to do it, performs an enormous work, and this, I am told by disinterested parties, is what Paul's barrel does. If coarser screens were used it would of course work faster, but the unprecedented fineness of the screens preferred by Mr. Paul in working gold ores, is justified by the increased yield of the ore. In treating silver ores this extreme degree of fineness might not be required; it is a point to be regulated by trial—that which is on the whole most profitable being the best.

I wish it to be distinctly understood, that in speaking of Paul's pulverizer, I do not mean Paul's dry amalgamation *process*, which, though it seems to be a good thing for gold, is not yet proved to be so for silver, as the two metals exist in a totally different condition in ores.

PULVERIZING BARREL.

[54.] A pulverizing barrel may be made of wood, not to discharge continuously like Paul's, which also feeds itself, but to work by charges. It is made like the amalgamating barrel, and lined in same way. The ore is coarsely crushed before being put in, and a quantity of small iron balls, or pieces of hard rock, are put in with it. A proper adjustment of the speed with which the barrel rotates, causes the ore and balls to roll over with a grinding action, without much wear of the barrel. A grating keyed in the bunghole of the barrel, when discharging, retains the balls, and allows the pulverized ore to come out. The barrel may be mounted on a wooden shaft passing clear through, and with broad bands of iron on the ends for journals, and friction rollers under them; or it may be arranged in any of the ways described for amalgamating barrels, omitting of course the hollow

SILVER ORES. 61

journals for steam, and the precautions regarding iron inside, both of which are needless.

KENDALL'S BATTERY.

[55.] Kendall's battery is another recent invention, designed for small operations, which, like Crocker's, is ready for use when bought. It can strike any required number of blows per minute, but has no springs. I should think it a good machine for small operations. [For cut of this Battery see next page.]

NOICES' PULVERIZER.

[56.] Noices' pulverizer is a queer looking machine, in which the rolling action of the Chili mill is combined with the grinding of the arrastra. It seems simple and efficient, is worked with water, and though designed especially for crushing and amalgamating gold rock, would work well in treating silver ore without roasting. From this machine I have derived the idea of a cheap rock breaker, or grinder, now to be described.

62 TESTING AND WORKING

KENDALL'S BATTERY.

A CHEAP ROCK BREAKER.

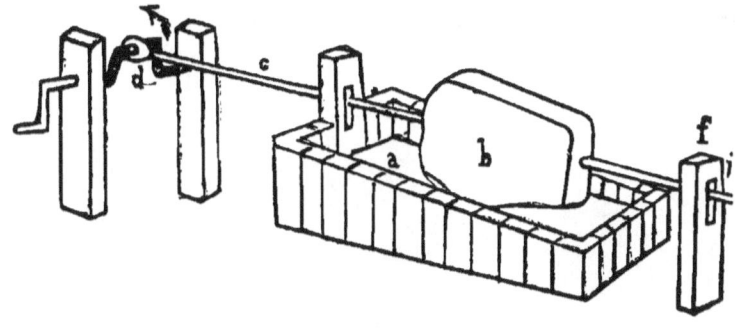

[57.] In the above cut, *a*, is an inclined table, made of one or more pieces of rock, enclosed by a low wall of rocks or wooden lags. On the table rests a mass of rock *b*, which is connected by the iron rod *c*, with the crank *d*; another iron rod *e*, also fastened into the rock *b*, passes through a slot in the post *f*. The ore to be broken is thrown upon the table in front of the moving rock, which, by the rotation of the crank in the direction of the arrow, is lifted and drawn forward over the ore. As the crank continues its rotation, the weight of the rock, and the sliding movement imparted to it, crush the ore beneath it, and cause it gradually to work out at the lower end of the table, when it may be transferred to a pulverizing barrel, or sifted, and the coarser part reground in the machine. This is merely a suggestion, and improvements will doubtless be made by intelligent miners.

AMALGAMATORS.

[58.] The man who, though poor in pocket, has rich ore, need not despair of getting along, if he has energy and some ingenuity. A five gallon beer barrel will make an admirable amalgamator for 50 pounds of ore, which, if worth $500 per ton,

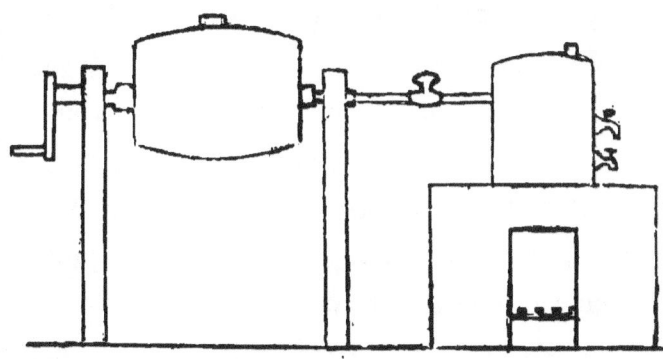

will yield as much money as a ton of Comstock ore at $12.50 per ton. A 10 gallon barrel will suffice to extract the same sum from 100 pounds of $250 ore, and either of these, or even a larger cask, may be mounted, as follows:

Find the centre of each head, and cut a hole, three inches square for the smallest barrel, and large in proportion for other sizes. Make a wooden shaft to fit tightly in the holes, and long enough to form journals at each end. On one end fix a crank, and in the center of the other bore a $\frac{3}{4}$ inch hole, far enough to reach inside the barrel when the shaft is in place, and connecting with a cross hole through the shaft, so as to communicate with the interior of the barrel. Put the shaft in its place through the barrel, and enlarge the bunghole, either round or square, and make a bung to fit. Mount the barrel on supports, just as you would a windlass. Next, get a tin boiler, have the cover soldered on, and an inch hole made in it, to pour water in with a funnel. Next get a piece of $\frac{1}{2}$-inch gas pipe with a stop-cock in the middle, and a flange made of a collar soldered on two inches from one end, and solder the other end into the boiler near the top; and it will be best to have two small faucets set in, one six inches from the bottom of the boiler, the other higher up, for try cocks to show how much water you have. The hole in the top must not be plugged, but an iron or lead weight must be used to close it; if the hole be one inch diameter, the weight may be from one to two pounds; this forms a safety valve. Mount the boiler, with a fire place below it, so that the steam pipe may enter the hole in the end of the barrel shaft as far as the flange. Get from 5 to 20 pounds of copper, in the form of balls, from $\frac{1}{2}$ to 1 inch in diameter, or little bars, or pieces of any kind, not too thin so as to break up when used; pieces of $\frac{1}{8}$-inch sheet are good for a very small operation. When the barrel is charged with ore, and a little water, let the steam from the boiler pass into it before putting in the bung, to drive out the air, then put the bung in securely, and tie a cord round the barrel to keep it in. Turn the barrel by hand or foot, by dog, squaw, chinaman, horse, water or steam power, and you have a perfect amalgamator for working Aaron's process.

[59.] The ore may be ground in any of the machines described for that purpose, or if very rich, even in a large hand mortar, with the pestle attached by a cord to a spring pole. A wash tub fitted with some simple kind of stirrer, will answer for a separator. Work your ore precisely as directed, using bluestone instead of dichloride of copper, unless you choose to make the dichloride in your porcelain-lined pot.

[60.] I have supposed, thus far, that you have already tested the ore, according to the directions given, and found it to yield satisfactorily; but the same barrel, without the boiler, will answer for working roasted ore, only instead of the pieces of copper, it is best to use scraps of iron, in quantity from 10 pounds upward, according to size of barrel. If the ore must be roasted, you may build with rocks and mud, a little furnace similar to that described hereafter, but simpler, and as each square foot of hearth will take about 10 pounds of ore, a hearth four feet square will be ample for something over 100 pounds. A barrel for Aaron's process, to contain $1\frac{1}{2}$ tons of ore, is $5\frac{1}{2}$ feet long by $4\frac{1}{2}$ feet diameter inside, and is mounted either on journals, or on rollers. The ordinary Freiberg barrel answers if the journals be made hollow to admit steam. In calculating the size of a barrel for working unroasted ore, I allow from 20 to 25 cubic feet for each ton of ore, and it must be remembered that, owing to the necessity of admitting steam at the center, the barrel can never be quite half filled, so that a barrel of 100 cubic feet capacity would contain about 2 or $2\frac{1}{2}$ tons.

The barrel is lined with blocks, cut from 3-inch plank, and so placed that the grain of the wood is vertical to the axis of the barrel. These blocks are cut at the saw mill, and before being put in are bevelled at the sides with a jack plane, to suit the circle of the barrel, but the backs need not be rounded, nor is any nicety required in the work, but the bevel should be too much, rather than too little. Mechanics are too apt to make a bad job of lining a barrel, simply because they do the work too well; instead of requiring close joints, it is better if the blocks are somewhat rough, and warped sideways, so that in bevelling them the warp should not be taken off with the plane, and in placing them they should be put with two bulging sides, or two hollow sides together, which leaves room for swelling without bursting the barrel, or making the lining bulge inward, leaving a space behind it to get full of amalgam. I have known $3,000 of amalgam to be taken from behind the lining of a single barrel, which was lined by a good mechanic; and this accumulation is very troublesome and injurious, by destroying the equilibrium of the machine. In putting in a lining, it is only necessary to nail one piece here and there, just to keep them in place till the circle is complete, when they key themselves, and when wet swell tight. When worn through, the lining is removed and a new one put in, for which purpose the barrel is furnished with a manhole in the head, which is closed with a door, just as in a steam boiler, but made of wood.

The staves are made of 3 x 4 scantling, not rounded outside or inside, the heads of 3-inch plank of the hardest pine. There are four hoops of 3 x $\frac{3}{8}$ inch iron, with lugs and bolts to set them tight. The lining is two inches thick, and before putting it in, about 20 pounds of roofing petroleum is melted, and poured into the barrel, into which steam is admitted by the hollow journal, and it is put in motion. The petroleum soaks into the staves, and makes the barrel steam tight; the lining is then put in. For the purpose of charging and discharging, there is a 6-inch bung hole, and a bung which, when in place, comes flush with the lining inside, and is secured by an iron hasp. A small plug in the end of the barrel serves for taking out a sample of pulp for examination when required.

SILVER ORES.

[61.] The copper used is in the form of balls, or small bars of not more than one pound weight. The balls are cast, and the bars are sometimes cut from long round cast bars, but flat pieces, either cast, or cut from ¼-inch rolled sheet, are to be preferred, as exposing more surface to a given weight. When the pieces become so small as to be with difficulty separated from the amalgam, they are first passed through the retort, to recover adhering quicksilver, and then remelted and cast. Rolled copper retains quicksilver on the surface only, but that which is cast sometimes becomes permeated throughout with it. The quantity of copper used varies with the richness of the ore, a great excess giving rise to a poor quality of bullion, from the fact that more is worn off than can be consumed in the chemical action. About 50 square feet of copper surface answers generally for a barrel of 1½ tons capacity.

[62.] One way of preventing the mechanical wear of the copper, is to use only so much quicksilver as to form a hard amalgam, which then adheres to the copper. This method was practiced in one of the mills at Benton, and was supposed to economize quicksilver, though I beg leave to doubt this, not on theoretical grounds alone, but because I did not find it so in my mill. The actual consumption of copper, in proportion to silver extracted, has not been determined.

[63.] There are three ways to use the quicksilver in this process: *Firstly*—at least one flask is used for each $100 of silver in the barrel. This gives a liquid amalgam, which comes out free from sand, and the copper balls are simply coated with quicksilver. *Secondly*—only enough is used to form hard amalgam, which sticks to the copper, as already stated, making it necessary to put the balls into a pot of quicksilver to get it off, and yielding an amalgam which is mixed with sand. *Thirdly*—so little is used that the copper balls come out free from quicksilver and amalgam, and the latter floats through the pulp in the form of gray flakes and pellets. The last is probably the best, as it leaves the surface of the copper free, so that the ore comes in direct contact with it, and the whole of the quicksilver comes out of the barrel with the pulp, but it requires the use of a wooden vessel, intermediate between the barrel and the separator, in which the floating amalgam may be gathered by means of an additional quantity of quicksilver, for as already remarked the iron bottom of the separator causes the production of copper amalgam.

[64.] I have used a barrel in which the copper was fixed, in the form of one inch bars, of which there were 20, extending the whole length of the barrel, and fixed at the distance of one-half inch from the staves. It depends on how the quicksilver is used whether the amalgam sticks to the bars or not. This barrel needs no lining, as the copper, being fixed, does not wear it. The entering a barrel to line it, to fix the copper, or to remove amalgam, is a disagreeable job, and when it must be done, considerable time is consumed in cooling it for the workman to enter. To avoid this, and other inconveniences of the barrel, I have invented a machine which I will describe after I have finished about barrels.

[65.] The following drawing represents the Freiberg barrel used for roasted ore at Palmetto, Nevada. By drilling through the journals, or having them cast larger and hollow, and lining with blocks as directed, this barrel will answer for working unroasted ore; its capacity without lining is one ton of ore.

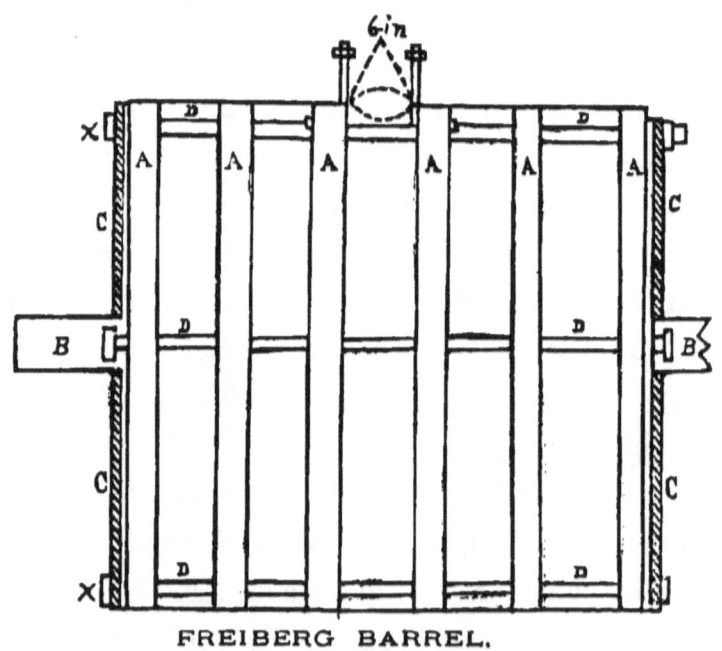

FREIBERG BARREL.

A, hoops. *B*, journal. *C* and *B*, iron flange. *D*, bolts.

[66.] A cheap barrel is made as follows: a 3-inch square bar of iron of the proper length, with journals six inches long turned on the ends, is bored at each end, precisely as directed for the wooden shaft of the small barrel. This iron shaft passes through the barrel, and is secured by cast flanges bolted to the barrel heads. The flanges are cast in two parts, with a strong hub through which the shaft passes, and the two sections are fastened together by means of a rib along each edge of the joint, through which bolts pass. This is not absolutely necessary, however. The square

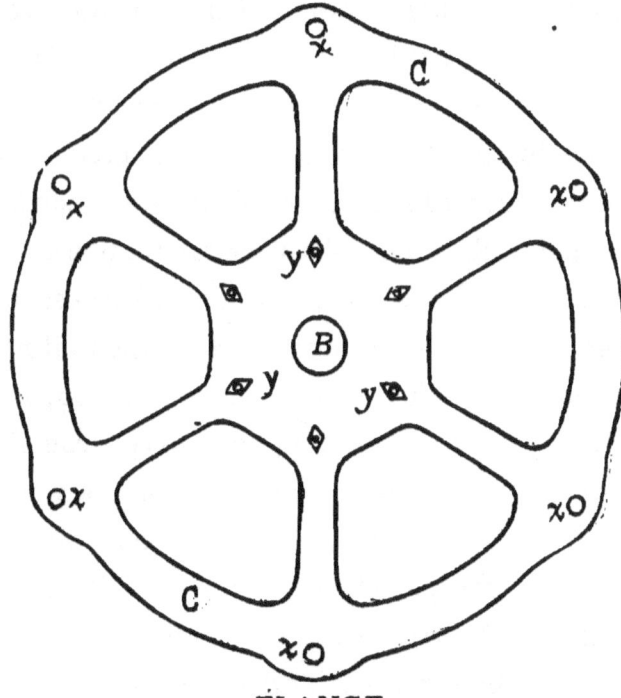

FLANGE.

x, bolt holes. *y*, holes for lag bolts running into head of barrel.

opening in the hub is slightly smaller than the shaft, and is so placed that the division of the two sections of the flange crosses it diagonally. The places for the flanges being accurately marked on the shaft, they are secured by a wrought iron band shrunk on to the hub. The barrel heads are then fastened to the flanges by bolts passed from within, with the nuts outside, and the heads supported by 3-inch countersunk cast washers, sunk flush into the wood. These are protected by a covering of putty, (better than white lead,) or a mixture of linseed oil and dust, and further by a wooden cap soaked in hot petroleum, with which the whole inside of the head is also painted. The shaft being mounted, the heads are turned true, the iron shaft encased with wood soaked with petroleum, and having in it holes corresponding to the cross holes in the iron, lined with short pieces of pipe tapped into the iron shaft to keep the steam from the wood; and a wooden cap, opening horizontally, is pinned on over each hole to keep the pulp from falling in The staves, which should be grooved and tongued for a lined barrel, are then put on, and nailed, to hold hold them until the bands are on and tightened with the draw bolts, after which the barrel is treated with hot petroleum and steam, as already described, and is then ready for lining. Some use a stuffing box in the end of the shaft where the steam pipe enters, but a simple flange on the pipe is sufficient. It is important that no iron work be left exposed inside the barrel, as the action of the chemicals would soon destroy it.

[67.] The trough or launder into which the barrel is discharged, and which conveys the pulp to the separator, is arranged in any convenient way, the best being that in use at Freiberg, as seen

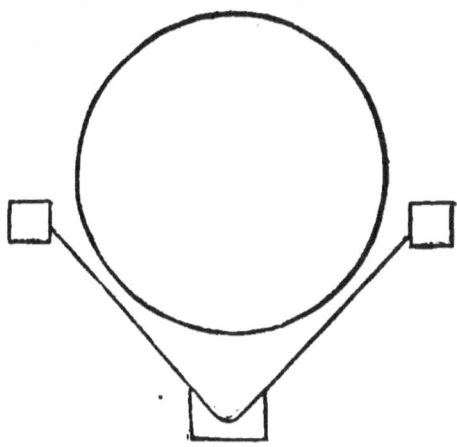

in the cut, the bottom being made of a solid timber, and the sides of matched boards, and extending under any number of barrels, mounted end to end. At some convenient part, between the barrels and the separator, is a pot to retain the quicksilver as already said.

78 TESTING AND WORKING

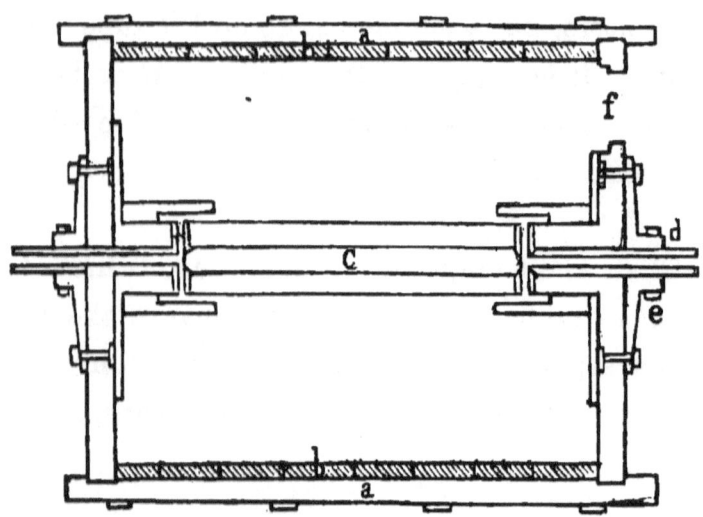

a, stave. *b*, lining. *c*, iron shaft. *d*, cast flange. *e*, wrought band. *f*, manhole. *g*, cap over steam hole.

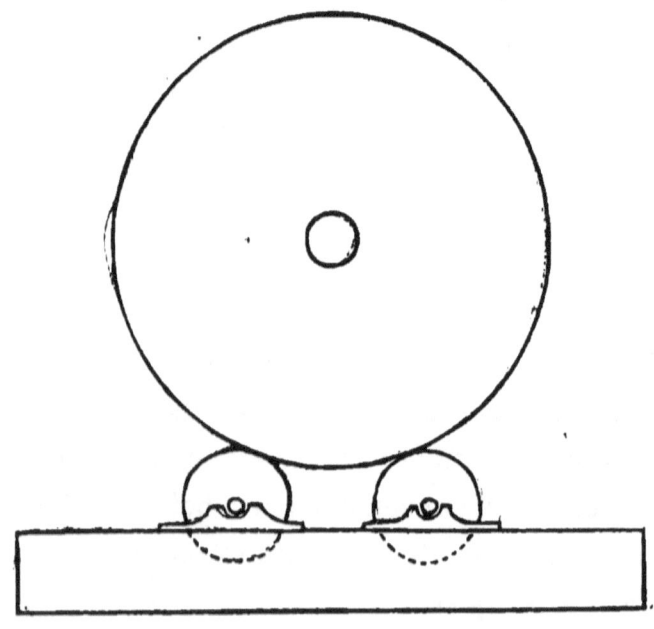

[68.] A barrel of any required capacity, even up to 50 tons, may be mounted on rollers as in the sketch. The rollers work on a cast, circular flanged rail, which surrounds the barrel. Mr. G. F. Deetken has built such a barrel, with which to work copper solutions, which is 12 feet long by 8 feet diameter; it is moved by a screw working into a circular rack surrounding the barrel. This barrel may be charged while in motion, through the circular opening in the end, steam being admitted at the other end.

AARON'S AMALGAMATOR.

[69.] Figure 1 is a perspective view of my device. Fig. 2 is a transverse vertical section.

A is a vessel of cylindrical or nearly cylindrical form, of which the lower portion is fixed and the upper portion is movable as a lid or cover, and within which is mounted a paddle wheel, B, with oblique floats $c\ c\ c$, and which is provided with openings $d\ d$, in each end, as shown. The wheel rotates in close proximity to the bottom of the vessel A, so as to pass through a mass of pulp, and dip in a bath of mercury contained therein.

I prefer to construct the vessel in an oval form with its longest diameter in a horizontal plane, so

Fig. 1.

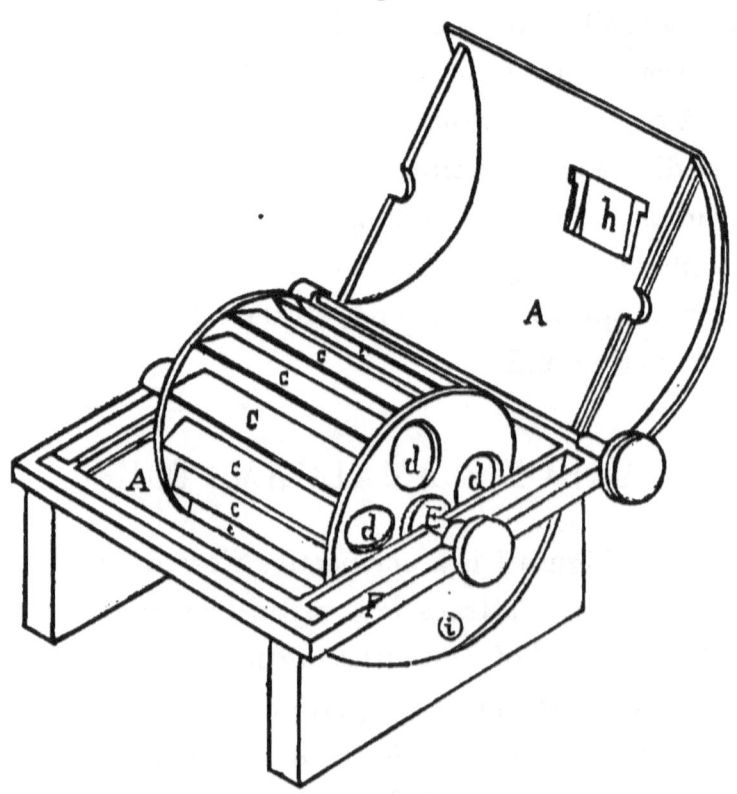

that a space will be left in front and rear of the wheel, while it moves close to the bottom and top of the wheel in its rotation.

The floats $c\,c\,c$, of the paddle-wheel may be made of any suitable material, but preferably of wood, faced with copper on the inner surfaces, and the wheel is secured upon a shaft, E, which bears upon opposite sides of a hinged frame, F. This frame, F, is so connected with the vessel, A, by means of its hinges as to admit of its being lifted at one end so as to raise the paddle-wheel out of the lower half of the vessel. The upper half or

Fig. 2.

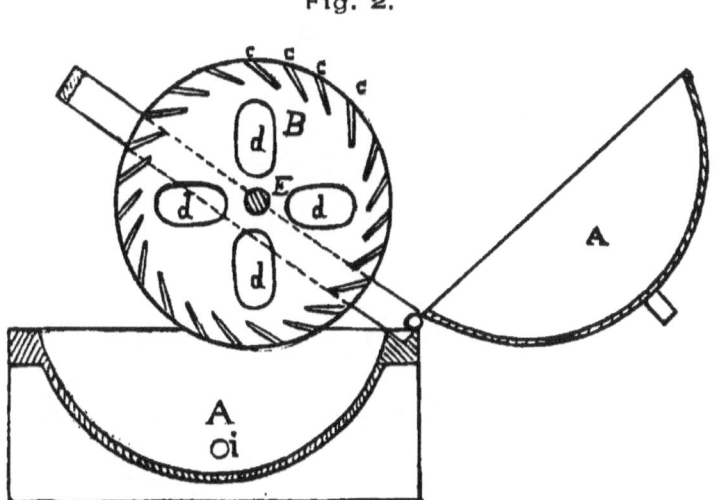

lid of the vessel A, is also hinged concentrically with the frame F, and may be raised together therewith, or separately. The vessel A, may be constructed of iron, or in cases where the presence of metallic iron is objectionable, of enameled iron, or of wood, or other suitable substance.

4*

Motion is given to the paddle-wheel B, by means of a belt and pulleys revolving concentrically with the frame, so that the rotation of the paddle wheel is not interfered with by raising it together with the frame.

The vessel, A, is provided with an opening, h, in the upper part or cover, and has a plug, i, in the lower part, to facilitate the introduction and discharge of the ore to be treated.

THE SEPARATOR.

[70.] The ordinary separator is not suitable for use in connection with barrels, especially when treating unroasted ore by Aaron's process. The

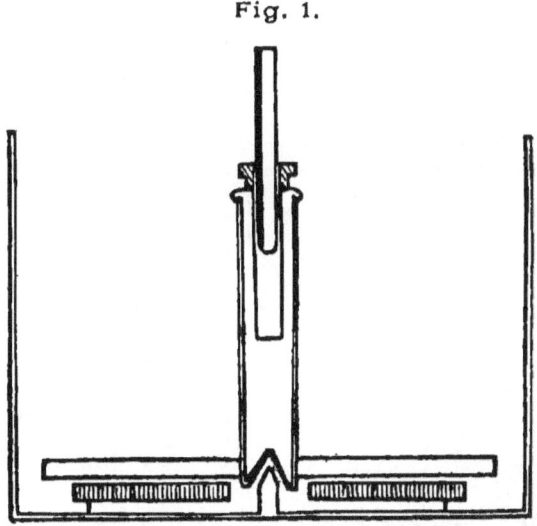

Fig. 1.

kind I prefer for this purpose is shown in **Fig. 1**, and consists simply of a wooden tub six feet in diameter and four feet deep, in the bottom of which is an iron annular disc, raised an inch and a half from the bottom of the tub by a smaller disc of wood under it, which leaves the iron disc projecting about two inches. The muller is made of scantling in the manner shown in Fig. 2, and the arms are short, not reaching to within six inches of the side of the tub.

Fig. 2.

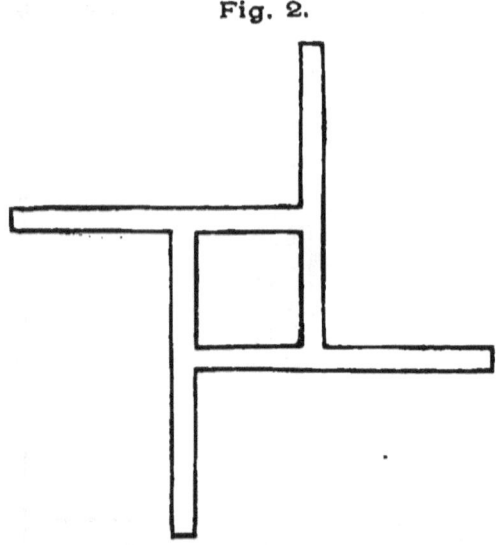

The mortises are arranged with keys, so that the muller can be clamped tightly to the square wooden shaft, the lower end of which runs on a low wooden cone having a core of mountain mahogany, and rising from the bottom of the tub up through the iron disc to a height of several inches. The upper end of the wooden shaft is connected with a short iron shaft, working in boxes attached to the framing, and carrying the pulley or gearing. The wooden shaft should be arranged to slide upon the iron portion, so that the muller may drop as it wears away, or may be raised with a lever in case of accident. The shorter the arms are the higher speed is used, and it should be such that the sand and particles of quicksilver are kept in continual suspension, till the latter, gathering together in a mass, settles into the annular space around the iron disc, where it remains undisturbed, while the sand remaining in suspension is drawn off with the water by successive plug-holes in the side of the tub. Any pieces of rock or copper, etc., which may accidentally find their way into the machine lodge under the projecting edge of the iron disc.

The pulp, as discharged still warm from the barrel, always contains copper dichloride in solution. By working it for an hour or two before adding water, in a separator with an iron bottom, which is scoured by a wooden muller, a copper amalgam is formed, thus preventing a large loss of quicksilver. It is for this reason that the bulk of fine amalgam from the barrel is not allowed to enter the separator, where it would become mingled with the base amalgam there produced.

RETORTS.

[71.] A small retort is improvised by cutting the top off a quicksilver flask, the cover being formed of the upper part of another flask, into which is screwed a piece of pipe about four feet long, curved near where it enters the cover, so as to incline downwards when the retort is in work. The cover is stretched on the horn of the anvil so as to overlap the body of the retort, and an iron band is slipped over the joint, which is well plastered with a stiff paste made of sifted ashes and water, with which also the inside of the retort is painted before putting the amalgam in. The pipe is wrapped with sacking, on which water drips continually from a barrel. The end of the pipe must not dip into the water in the receiving pot, as there is danger of an explosion from the water being drawn up the pipe into the retort. A retort should not be more than two-thirds filled if the amalgam is fine, nor more than half when much base metal is present. It is a good plan to have a piece of stout wire or nailrod passing through the

ROASTING FURNACE.

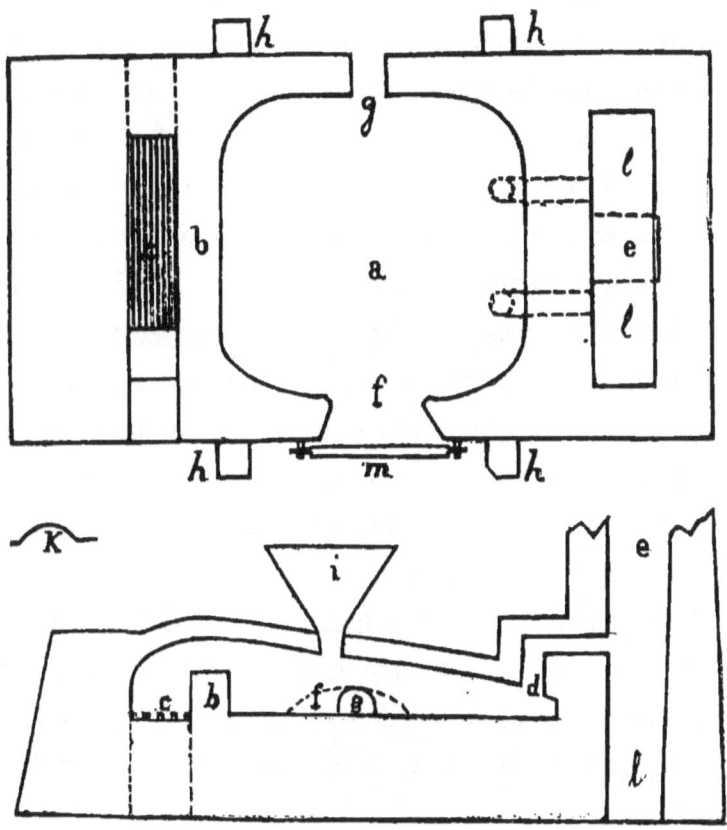

a, hearth. *b*, bridge. *c*, fire-place. *d*, flue. *e*, stack. *f*, working door. *g*, discharging door. *h*, binder. *i*, hopper. *k*, iron to support working door. *l*, a dust chamber into which the flues open, and which being drawn into a square above the flues becomes the stack, *e*. *m*, roller.

pipe into the retort, with which the operator may assist in the expulsion of any dirt which may pass over and obstruct the pipe. The fire is made on the top of the retort first, and the operation must never be hurried, but allowed to proceed gently till nearly all the quicksilver has passed over, when the heat may be raised to a dark red, and so kept till no more quicksilver can be got. The retorted spongy silver may be sent to the mint, without being first melted into bars. Explosions only occur through ignorance or carelessness. If there is good reason to think that the pipe is stopped, as sometimes happens through the swelling of the amalgam, when overcharged, or if, as sometimes happens with cast retorts, quicksilver is seen escaping from a crack, the fire must instantly be drawn, or the operator must emigrate.

For quantities of amalgam up to 200 pounds, cast iron pot retorts are used; beyond that, the cylindrical or D patterns, permanently set in masonry, and provided with a water jacket for the pipe.

[72.] The hoes for stirring the ore are 16 feet long, and in order to use them with convenience, an iron roller resting on bearings built into the brickwork, is placed before the working door, a little above the level of the hearth. The shank or handle of the hoe is best made of gas pipe, all but about five feet next the head, which is solid. The head or blade is either of cast or wrought iron, and as they do not last long, several spare ones are kept on hand. There must always be several hoes in readiness for use, so that they may be changed when too hot. A fire hook is also necessary for trimming the fire, and some use long-handled iron spades for moving the ore from end to end of the furnace, as must be done several times in order that all may be equally heated, but workmen soon learn to do this with the hoe. Rakes with cast iron heads are also sometimes used, as they are more easily drawn through the ore than hoes, but the hoe cannot be dispensed with, and in some mills nothing else is used.

The working door of the furnace is kept open nearly all the time when roasting, but is closed when required, by a piece of boiler iron, cut to fit the opening, and furnished with two small holes, about four inches apart, to receive the prongs of an iron fork with which to lift it; when in place it rests against the iron strap, K, which is built into the opening, and forms a rabbet. The fire door, and that at the back, are also of boiler iron, hung upon pintles set in the masonry. The ashpit opens on the opposite side from the fire door, which causes the heat to be more even in the fire place. The dust chamber should be as large as possible. It is a good plan to carry the flue under the drier in front of the battery, forming a dust-chamber, and using the waste heat from the furnace to dry the ore for crushing. The draught of a furnace must not be too strong, but just enough to keep the fumes from coming out of the open door. To regulate the draught a damper is provided, which may be conveniently placed in the stack, being simply a piece of sheet iron turning on a pivot, just like the damper in a stove pipe.

[73.] The main body of the furnace may be built of rocks and mud, the hearth, or sole, walls and arch, of brick; those in the sole being laid on a bed of sand. The bricks in the arch are laid end up on a form made of moist sand, which is drawn out as soon as the work is done, to let the arch settle as it dries, for which reason, also, it is built with a little more spring than is wanted. In addition to the longitudinal curve, it is slightly arched transversely. In the walls every alternate course should be headers (bricks laid crosswise), and they must be supported by binders, formed of strong posts set into the ground, and united at the top by an iron rod, or a tie-beam, across and above the furnace. If the ends of the furnace, forming abutments to the arch, are built very heavy, as represented, it is not necessary to bind it lengthwise. Adobes will answer for building a furnace, though they will not last so long as bricks. The bridge, being exposed to great heat and hard knocks, is best made of one or two pieces of fire stone, or of firebrick mixture beaten, while moist, into a mould; but common bricks answer pretty well.

[74.] In a much smaller furnace the walls may be made of a concrete of mud and small rocks, and the arch of adobe material, laid, like plaster, on a form of damp sand covered with paper. The plaster is laid on two inches thick, and as it will crack some in drying, it is cut through with the trowel in squares; before it is quite dry another layer is put on, and so on till it is thick enough, the supporting sand being drawn out as soon as the arch will stand.

[75.] I have given these details about the furnace for the benefit of those whose situation makes it necessary to make a start with as little cash outlay as possible; but, as soon as circumstances will allow, I strongly advise the use of a revolving cylinder—that of Brückner being the best for a small operation.

AARON'S LEACHING APPARATUS, No. 1.

[Patent to be applied for.]

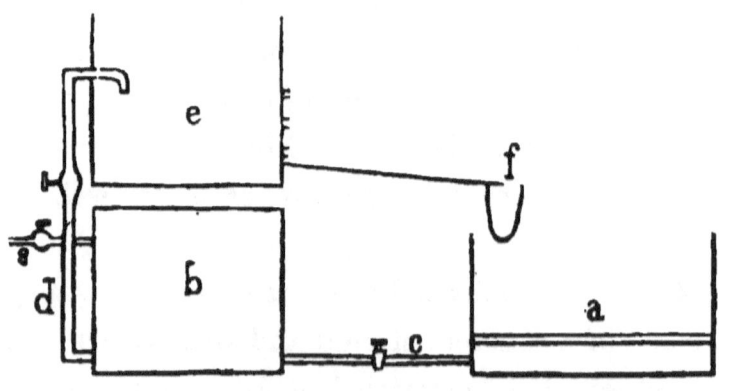

a is an ordinary leaching vat; b is a wooden vacuum chamber; c and d are pipes, either of wood or rubber, furnished with stop-cocks; e is a precipitating vat; f a flannel bag; and g a steam pipe connecting with a boiler. The ore being placed in a, and saturated with the leaching fluid, (the cock in c being closed) the chamber b is filled with steam,

SILVER ORES. 93

the air being expelled through d which is open; d is then closed, and steam shut off, when c being opened, a vacuum is formed in b by condensation of the steam with which it was filled. The pressure of the atmosphere on the surface of the fluid in a, forces it rapidly through the bed of ore into b; the cock c is closed, that in d opened, and steam again admitted into b, and by its pressure forces the solution through d, into e, where it is treated with the precipitating agent, and when clear, is drawn off and passed again to a, through the flannel bag f, by which any floating particles of precipitate are retained. The bulk of precipitate remaining in c, is then removed.

[76.] This is the simplest form of the apparatus, and is given merely for the purpose of explanation.

The following is a better arrangement—

1, 2, 3, 4, are leaching vats; a is the vacuum chamber. No. 1 being charged with ore, and the leaching solution, No. 2 is also filled with ore, and the solution being drawn from 1, by means of the vacuum chamber, is thrown into 2, while 1 is supplied with a fresh lot of the solution from the precipitating vat, not represented as it is above *a*. In like manner the solution from 2 is passed to 3, that from 1 to 2 again; from 3 to 4; again from 2 to 3 and 1 to 2; then from 4 to the precipitating vat, from 3 to 4, 2 to 3 and 1 to 2 again. The silver or other metal of 1 will by this time be exhausted, and it is emptied, and recharged with fresh ore,

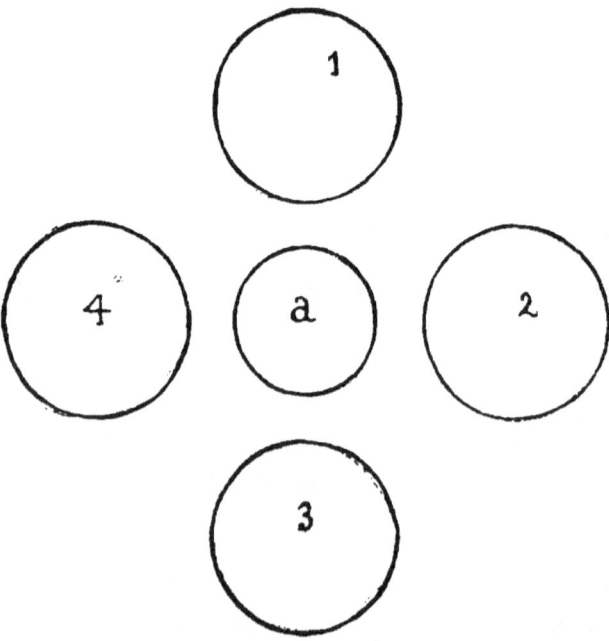

and the next discharge from 4 is passed into it, and from it to the precipitator, so that it now becomes the fourth in the series, the former No. 2 becoming No. 1; and so the operation goes on, the last vat charged with fresh ore being always No. 4 of the series, and the solution from it being passed to the precipitator saturated with metal. In this way a concentrated solution of metal is always had for precipitation, the metal is dissolved rapidly out of the ore by the constant movement of the dissolving solution, and the finished ore is quickly drained.

[77.] If the vacuum vat is divided vertically into four compartments, by cross partitions not extending to the top, all four vats can be drained at once, and the fluid thus passing more continuously the metal will be dissolved still more rapidly, for it is well known that sugar, or any other soluble substance, is dissolved sooner if stirred in water than if left at rest; and still sooner if placed on a filter and fresh water constantly passed through; so, in like manner the metal in the first vat is quickly dissolved by the constant flow of fresh liquid, while the inconvenience of operating on the weak solution which comes from it towards the last, is obviated by passing it through the other vats, and lastly through one containing a fresh charge of ore, where it becomes saturated. In case of ores requiring several distinct leachings, or washing before leaching, there may be two or more sets of vats, mounted on trucks, and running on tramways; also, as many vacuum chambers. The vats then replace each other in the respective sets, and when one is finished it is wheeled to a dumping place, emptied, and after being recharged, takes its place in the first set. Large stationary vats may be built with the filters inclined, and the exhausted material washed out with a stream of water from a hose. Roasted silver ores are washed before leaching for silver, by passing through them acidulated water, obtained by causing the fumes from the roasting furnace to pass through, or in contact with water, either by means of a Cagniardell's condenser, or a blower, or a stack with shelves, rocks or coke in it, which are kept wet by a stream of water.

[78.] APPARATUS, No. 2—(Patent to be applied for.) I have devised another apparatus, in which the dissolving fluid is caused to circulate continuously, either through the same vat, or through a series, by means of centrifugal force. This may be used for extraction of gold with chlorinated water, or of gold and silver, together, with chlorinated brine. The following diagram shows its operation:

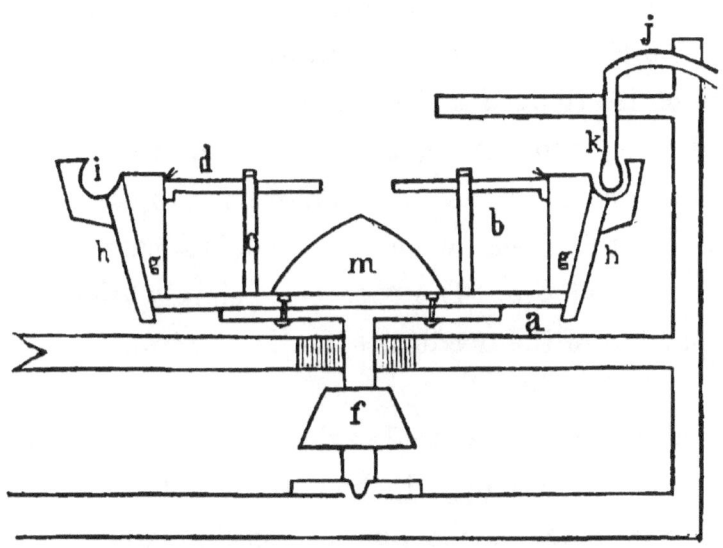

[79.] The leaching tub is securely bolted to the revolving iron table a; the ore is seen at b, occupying the annular chamber formed by the circular partition c, and cover d. The partition c is constructed of perforated wood, and canvas, so that the dissolving fluid, being poured into the central opening, and the tub being made to rotate rapidly by a belt working on the cone pulley f, it is driven by centrifugal force through the ore b, into channels between the blocks g, and reaching the sloping side of the tub *h*, flows upward and into the channel *i*, which is attached to, and surrounds the tub. A fixed scoop and pipe *k*, dips into this revolving channel, and the fluid, charged with metal, is forced up through the pipe *k*, to which is attached a piece of rubber tube *j*, and is delivered into the center of another similar machine, or a precipitating tank, or back to the center of the tub from which it came, to pass again through the ore. The cone *m*, is simply to occupy a portion of the needless space in the tub.

[80.] When the ore is exhausted and drained, the machine is stopped, the cover *d*, removed, and the tub being again put in motion the ore is scooped out of the annular chamber by a fixed scoop not shown in the diagram. The speed of rotation is regulated as required, by means of the cone pulley.

[81.] The sketch represents a mill for working two or three hundred pounds of rich ore per day. The large wheel may be a water wheel, or treadmill; the pulverizing barrel *a*, is a beer or wine cask, as is also the amalgamating barrel *b;* the separator *f*, is a large washtub. Beneath the pulverizing barrel is a seive *c*, to retain the balls or lumps of rock, the ore falling into the box *d*. Under the amalgamating barrel, which is driven by a belt

A VERY SMALL MILL.

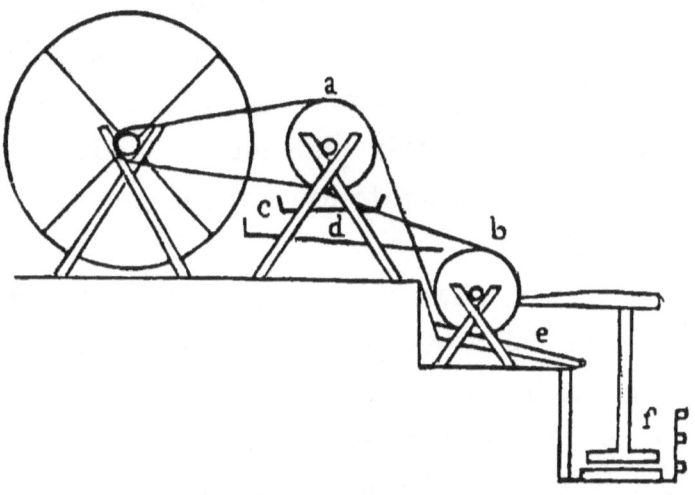

from *a*, is a trough *e*, to carry the pulp to the separator, which is worked by a half-twisted belt from a pulley on the shaft of *b*. The boiler for heating the barrel is not represented. In a larger affair the pulverizer *a*, may be replaced by a small arrastra.

Perhaps the best way to utilize a very small

power, is to apply it all to one part of the process at a time. Thus, ore may be crushed for several days in succession, and then amalgamated in one day. In this way there are not so many things to attend to at once, which is one of the difficulties of a small mill; of course, however, the machinery to work a given quautity of ore per day must be larger than when it is all used at once, and each machine will be idle a part of the time. However, where only two or three tons of ore are worked per day, this plan must be adopted to some extent in order to use a roasting cylinder to advantage.

The power of two chinamen, working turn about, will work 200 or 300 pounds of ore a day; that of two common horses, 400 to 500 pounds, and 60 miners' inches of water, or about 100 cubic feet per minute, on a 30-foot overshot wheel, with a small battery or an arrastra, will work from two to three tons in 24 hours.

[82.] An error into which inexperienced persons are apt to fall, on first starting a mill, is being needlessly alarmed at the small yield of amalgam, and the disappearance of a quantity of quicksilver, owing to the retention of both in the various vessels through which it passes. Positive results can only be arrived at, after a run of several days or weeks, by a thorough clean up, when the dies of pans must be lifted, and the lining of barrels removed. An examination of the tailings, by panning, will show whether much quicksilver is being lost mechanically; and as a chemical loss is usually accompanied by granulation, and consequent mechanical loss, or at least a difficulty in "settling," and the appearance of white or brown powder in the tailrace, a careful examination will afford a good indication of the actual loss. An assay of a true sample of the tailings compared with the value of the ore, will at any time indicate the percentage to which the ore is being worked. The best way I know of to take a sample of tailings, is to take a little pulp from each pan or barrel, as it is being discharged, then, after thoroughly mixing the general sample, take a small portion in a saucer, and carefully pan it into a vessel of water, retaining all the amalgam in the saucer. After the water has settled clear, pour it off as far as possible without losing any earthy matter, dry the residue and assay it. The result will at least not be too high, because the sample being taken before the pulp entered the separator, it will be richer than the final tailings by as much as is extracted by the separator. However the error is on the right side, for no one who thinks he is doing well will grumble at finding that he is doing better than he supposed.

[83.] I will now describe an invention which I have made, to replace the settling tank in a wet-crushing mill, which will obviate twice handling the ore, once to remove it from the tank, and again to put it into the pans. This apparatus has also the advantage of retaining all the slimes, together with the sandy portion of the pulp, a point which millmen will appreciate, and it may, in addition, be arranged so as to be used as a preparing vessel, in which to treat the ore with chemicals.

a is a vat furnished with a stirrer b, and a false bottom c, which forms a filter. The pulp from the battery is delivered into a by the trough d, and by the action of the stirrer is prevented from settling. The water passing through the filter c, fills the pipe e, the lower end of which dips into a vessel of water g. The cock f, being now opened, a suction is established through the pipe by the weight of the column of water within, and the surplus water in the pulp is rapidly drawn off, the flow from the battery being diverted at the proper time into another similar machine. When the pulp in *a* is of the proper consistency, the cock *f*, is closed, the plughole *h*, opened, and the contents of the

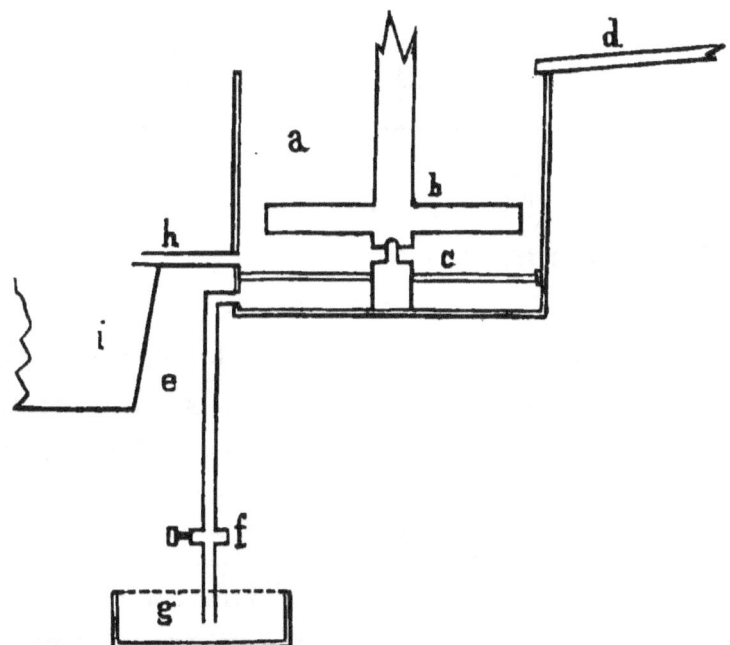

vat discharged into the pan *i*. Previous to this, however, the pulp in *a* may, if desired, be treated with chemicals and steam, to prepare it for amalgamation. The machine may of course be geared from below li

In large mills the vat or tank would be in the usual form, with two or more stirrers, as shown in Fig. 2.

FIG. 2.

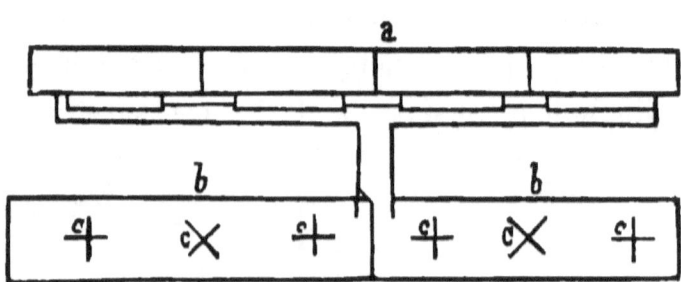

a is a row of four batteries, with sluices leading into either of the vats *b b*, each of which has three stirrers, *c c c*. The entire bottom of each vat, except the rests for the stirrers, is a filter, and each vat holds enough pulp to charge all the pans.

The vats may be arranged for upward filtration, if desired, as in Fig. 3, in which the letters correspond with those in Fig. 1.

The filter *c*, in this arrangement, can be easily cleaned if it should become choked, as, by turning the pipe *e*, in the side of the vat, its position can be reversed. It may also be arranged around the sides of the vat. A vat with a stirrer or stirrers may be used with or without a filter. In the latter case, the stirrer is best made in the form of an obtuse cone, with the apex downward, thus, ⊥, in order that it may easily work its way into the sediment, and it is arranged to be raised or lowered at pleasure; the pulp is allowed to settle, and the

FIG. 3.

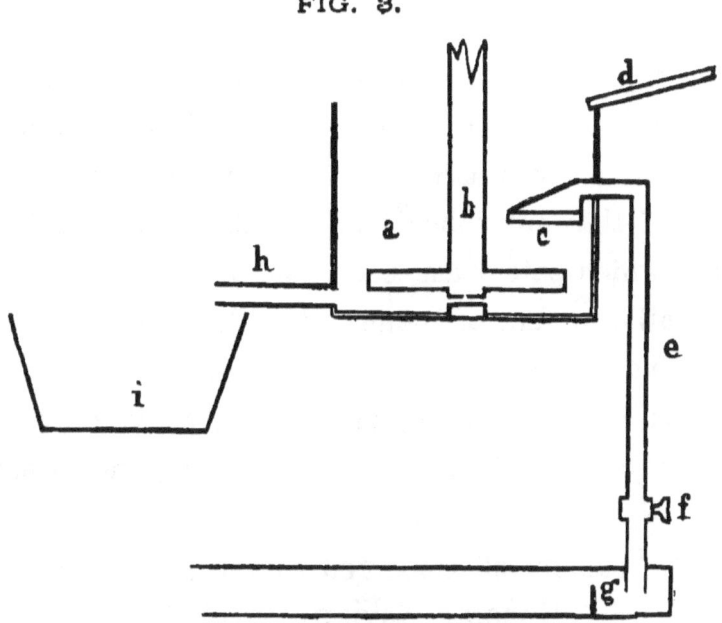

surplus water to flow out, by gates or plugs, just as in the common ore tank, and when a sufficient quantity of ore has entered, the incoming stream of pulp is stopped, and the water being reduced by the plug-holes to the required quantity, the stirrer is put in motion, and gradually lowered to

5*

stir up the sediment, so that, forming with the water a semi-liquid mass, or pulp, it may flow through the spout h into the amalgamator, or it may even be ground or amalgamated in the vessel a, if said vessel is so constructed as to answer these purposes.

[84.] NOTE.—In case persons desiring to use dichloride of copper should be deterred from so doing by the fact of a patent having been granted on the use of that substance "in barrels, pans or tubs," and which patent has been sold to certain parties for a large sum of money, the following from the *Mining and Scientific Press*, written by Dr. Lanszweert, may be useful, antedating, as it does by some years, and therefore invalidating, the patent:

The Problem of Gold and Silver Extraction—A General Review. No. 16.

BY PROF. L. LANSZWEERT.

In continuing my remarks upon silver processes, the last of which appeared in your issue of July 29th, I would say that in the Freiburg process, it is of the utmost importance that no free protochloride of copper should be present in the reduction of silver ores, as by giving up to the mercury one-half of its chlorine, a considerable loss of that metal is produced by chemical reaction. When this does occur, lime is added for the purpose of decomposing the excess of chloride of copper, and its prejudicial action on the mercury is arrested.

USE OF DICHLORIDE OF COPPER IN THE PAN PROCESS.

To avoid this loss, which, at the end of the year forms a considerable item in the expenses of a mill, and to facilitate a more direct and easy decomposition of the sulphurets of silver in the Washoe ores (admitting the *rationale* of the theory of the patio process as explained in my last number), I availed myself, during my stay in Washoe, of the opportunity to try the direct use of dichloride of copper ($cu^2 cl.$) on the different ores there presented. The success realized quite surpassed my expectations. The results were obtained by comparative experiments made on a large scale, by reducing from five to five hundred tons of the various Washoe ores, such as Gould & Curry No. 2, Chollar, Potosi, Olney, Rogers, Gold Hill, etc., and were attested by certificates from Guido Kustel, Col. Breevort, Mr. Lambert, J. Corey, Joseph Trench and others, as the experiments were conducted at their several mills. They averaged, according to the nature of the ores, from within five to twenty per cent. of the assay value of the ores.

The quantity of dichloride of copper used was based on the analysis of the ores, and the percentage of the sulphide of silver they contained. Taking, for example, an ore assaying $100 per ton, we would have, according to analysis, sixty-two ounces of sulphide of silver, equal to fifty-four ounces of metallic silver, at $1.29 per ounce, which is worth $70, and free silver at $30 = $100.

The patio process requires for the reduction of the above amount of sulphide of silver 125 ounces of sulphate of copper, 590 ounces of salt, and 100 ounces of mercury; the mercury being lost by decomposition as a chloride of mercury—the calomel of the apothecary.

Now, if we use, instead of the above, the *dichloride* of copper, 47 ounces (the quantity required for the reduction for the amount of sulphide of silver present), plus 400 ounces of salt and 100 ounces of mercury, there will be no loss of the mercury.

In nearly all enterprises of this kind a great many obstacles are thrown in the way of success, by reason of the stubbornness, incapacity or avidity of those having charge of the different works. So in this case; notwithstanding the proof was abundant of more favorable results than were being obtained by the old processes, still such was the systematic opposition presented by superintendents, amalgamators and others interested in retaining present positions, or in continuing the use of other processes, the writer found it impossible to introduce his process into permanent use in any of the numerous mills in Washoe. Time and circumstances will eventually accomplish for the mining interest of that locality, what the stubbornness or incapacity of the early operators there refused to allow; although thousands of dollars might thereby have been saved to individual companies alone.

Although the preparation of dichloride of copper is very easy and simple, still a great many attempts to prepare it have failed; the operators preparing instead, a mixture containing a large percentage of protochloride and very little dichloride of copper; and being in consequence disappointed in the results from the use of this preparation, they have condemned a process which my own careful experiments on a working scale have proven to be the best yet adopted for the reduction, without roasting, of a certain class of Washoe ores.

TABLE OF CONTENTS.

TESTING ORES FOR SILVER.

Subject.	Section.	Page.
Rock Formation	1	7
Test for Silver	2	8
Silver, Indication of Presence	3	10
Silver Test with Heat and Water	4	11
Silver Test with Nitric Acid	5	12
Silver Test with Blow-pipe	6	13

TESTING FOR A PROCESS.

Extent and Richness of Ore	7	14
Smelting Ores	8	15
Selecting Sample	9	16
Working Sample	10	16
Copper Test	11	18
Bluestone	12	19
Appliances for Testing	13	20
Roasting	14	21

WORKING ORES.

Aaron's Process	15	23
Superheated Steam	18	25

TABLE OF CONTENTS.

Subject.	Section.	Page.
Dichloride of Copper, Preparation	19	26
Protochloride of Copper	20	26
Carbonate of Copper Ores	21	27
Use of Copper and Iron	22	29
Quantity of Chemicals	23	29
Carbonate of Lime	24	30
Class of Ores Worked by Aaron's Process	25	30
Chloride Ores	26	30
Amalgam	27	31
Patchen's Process	28	33

ROASTED ORES.

Working Roasted Ores	29	34
Base Metals	30	35
Chloridizing Roasting	31	36
Directions for Roasting	32	37
Stirring	36	40
Heat	37	40
Want of Sulphur	38	41

LEACHING PROCESSES.

Classes of Leaching Processes	39	41
Smelting	40	42
Mexican Process	42	43
Krœncke's Process	43	44
Chilian Process	44	48

PULVERIZING MACHINES.

Subject.	Section.	Page.
Arrastra	46	49
Operation of Arrastra	47	50
Feed and Speed	48	51
Dry Grinding of Arrastra	49	52
Stamp Batteries	50	53
Screens	51	53
Crocker's Trip Hammer Battery	52	55
Paul's Pulverizing Barrel	53	57
Pulverizing Barrel	54	60
Kendall's Battery	55	61
Noice's Pulverizer	56	61
Cheap Rock Breaker	57	63

AMALGAMATORS.

Subject.	Section.	Page.
Cheap Amalgamator	58	65
Grinding the Ore	59	67
Directions for Making a Barrel	60	68
Copper	61	71
Preventing Mechanical Wear	62	71
Using Quicksilver	63	72
Copper in Bars	64	73
Freiberg Barrel	65	73
Cheap Barrel	66	74
Trough	67	77
Barrel on Rollers	68	79
Aaron's Amalgamator	69	79
Separator	70	83

RETORTS.

Subject.	Section.	Page.
Improvised Retort	71	86
Roasting Furnace	72	88
Furnace Tools	72	88
Furnace Building	73	90
Smaller Furnace	74	91

MISCELLANEOUS.

Aaron's Leaching Apparatus	75	92
Another Arrangement	76	93
Apparatus No. 2	78	96
A Small Mill	81	99
Sampling Tailings	82	101
Settling Tanks	83	102
Dichloride of Copper	84	106